- antennae generally clubbed — ...erflies, pp. 161 - 92
- antennae not clubbed — Moths, pp. 161 - 81
- narrow, without veins fringed with long hairs. Insects less than 6 mm (¼") long — Thrips, pp. 70 - 1
  - wings obviously hairy, hind wings much larger than forewings. Up to 40 mm (1½") long — Caddis Flies, pp. 160 - 1
- covering all or most of hind-wings when folded — Beetles, pp. 80 - 127
- mouth formed into piercing and sucking proboscis — Bugs, pp. 49 - 69
- forceps on end of abdomen — Earwigs, pp. 25 - 6
- no forceps
  - hind legs
    - long
      - developed for jumping — Grasshoppers and Crickets, pp. 28 - 38
      - thin for walking — Stick Insects, pp. 39 - 42
        - front legs thick to catch prey — Mantids, pp. 20 - 4
    - not long — Cockroaches, pp. 10 - 13
- head large — Dragonflies and Damselflies, pp. 5 - 7
- head small
  - hind-wings wider than forewings three segments on tip of leg — Stoneflies, pp. 27 - 8
    - base of hind-wings fold like fan, five segments on top of leg — Alderflies and Dobsonflies, pp. 72 - 3
    - no fanlike area — Lacewings, pp. 73 - 80
  - mouth formed into a beak — Scorpion Flies, pp. 129 - 32
- two or three long tails — Mayflies, pp. 4 - 5
- front and hind-wings joined together in flight by minute hooks
  - no waist present — Sawfly, pp. 193 - 6
  - waist present — Bees, Wasps and Ants, pp. 193 - 213
- front wings reduced to small clubbed antennae — Stylops, pp. 128 - 9
- hind-wings reduced to slender clubbed stalks — True Flies, pp. 135 - 59

See inside back cover

Insects of Australia

John Goode

INSECTS OF AUSTRALIA

With illustrations from the classic

THE INSECTS OF AUSTRALIA AND NEW ZEALAND

by R. J. Tillyard

ANGUS & ROBERTSON PUBLISHERS

ANGUS & ROBERTSON PUBLISHERS

Unit 4, Eden Park, 31 Waterloo Road,
North Ryde, NSW, Australia 2113, and
16 Golden Square, London W1R 4BN,
United Kingdom

First published in Australia
by Angus & Robertson Publishers in 1980
Reprinted 1984, 1987

National Library of Australia
Cataloguing-in-publication data.

Goode, John

Insects of Australia.

ISBN 0 207 13427 8.

1. Insects — Australian. 2. Insects — Identification — Handbooks, manuals; etc. I. Title.

595.7'0994

Printed in Hong Kong

ACKNOWLEDGMENTS

My thanks to Mr Don Kells and Ms Jenny Forse of the Zoology Department, Monash University, for their valuable help in the preparation of the manuscript. A difficult job of typing was magnificently handled by Ms Judy O'Neill and Ms Sue Rodda, who coped not only with a multitude of scientific names, but also with the author's corrections in indecipherable handwriting.

THE ILLUSTRATIONS

The illustrations in this book are taken from R. J. Tillyard's classic work *The Insects of Australia and New Zealand*, first published in 1926 and now out of print. They are outstanding studies, as informative, interesting and beautiful as when they were first published many years ago. They include eight magnificent colour plates, painted by P. Tillyard, the wife of R. J. Tillyard, over a hundred black and white paintings, also by P. Tillyard, black and white photographs by P. Davies, and over a hundred intricate line drawings by R. J. and P. Tillyard, A. Tonnoir and E. S. Gowlay.

In a few cases, Tillyard's book did not supply the required illustration. The extra line drawings have been supplied by David Mitchelhill.

Contents

Introduction

Apart from the ocean masses, the whole of the earth's surface is populated by insects. Entomology, the study of these incredibly varied and fascinating creatures, is an increasingly popular hobby and a rewarding pursuit.

Australia is rich in insect life and, especially in the warmer areas, has many species that are unique in size, colour and profusion. This book is intended to make it easy for the young collector or field naturalist to determine what insect has been found, and should provide the necessary data to enable species to be located in their natural habitats.

The number of insects identified to date in Australia (and many more are known but as yet undescribed), make it impossible for a book such as this to be more than an elementary guide. There are over 10 000 described species of Australian beetles, grouped in more than 100 families; obviously the selection of those families included here has had to be restricted. Basically, the aim has been to include all the common insects, together with those considered to be of particular interest, owing to their unusual or unique characteristics.

The text provides most of the basic information pertaining to each insect Order and, in some cases, even to a particular group or species. In conjunction with the simplified identification keys (which have been designed for the amateur naturalist and at times diverge deliberately from accepted scientific lines), the text should provide sufficient information for a collector to proceed at least part of the way towards identifying *any* Australian insect.

In order that the greatest possible space might be devoted to individual groups, the usual preliminary sections on structure, biology, collection and natural history have been omitted. There are many first-class books on these subjects available, and recommended works will be found listed on page 241.

If this guide can develop an awareness in people of all ages of the insects around them, and of the fact that many of these insects are not pests, but are in fact beneficial to man, it may play some part in helping people to understand their own ecology; to appreciate, for example, that the indiscriminate use of pesticides might well be reconsidered in the light of possible damage to insect species whose value to man could in the long run outweigh economic or other short-term considerations.

Wingless Insects

Subclass APTERYGOTA (wingless)

Although various insects lack wings, the only ones to be recognised as 'primitive' (never having developed wings) are the archaeognatha and silverfish.

ARCHAEOGNATHA (no common name)

Order ARCHAEOGNATHA

Archaeognatha resemble silverfish but can 'jump' by flexing the abdomen. They are found mainly under tree bark, among debris and in rock crevices.

The body is long and oval and tapers at the back to end in three slender tail appendages *(cerci)*. Archaeognatha have long antennae with many segments. These are longer than the body and considerably longer than a silverfish's antennae.

Archaeognatha are nocturnal vegetarians, living on algae, lichens and rotting vegetation. Although they are able to run swiftly, they normally escape predators by their phenomenal jumping, which can take them some 100mm in a single bound. They lay soft, orange-coloured eggs, about 15 to a clutch, which they deposit in crevices or in holes bored by their long ovipositor. The larvae pass through eight or nine moults and take from six months to two years to reach maturity.

There are only three described Australian species, representing one of the two families and 250 species known (including some extinct). *Nesomachilis australicus* is found in the rain forests of Queensland and New South Wales.

The cliff-dwelling *Allomachilis froggatti* grows to a length of 18mm and is found on the eastern, south-eastern and south-western Australian coasts. The Tasmanian *Machiloides hickmani* is found in grass clumps on sand dunes and beaches. A relative of the last species has been found near Canberra while others, as yet unidentified, have been found in forests in Victoria.

SILVERFISH

Order THYSANURA (fringe-tailed)

Like archaeognatha, the Australian representation of silverfish is small, with 23 described species in 3 families being only a small fraction of the 5 families and 330 species recorded.

Silverfish are active, fast-running, pointed insects with soft silver-grey oval bodies which taper to three tail appendages. They are covered with loose, easily-detached scales.

Spiders predate on adult and young silverfish. Some species eat paper and glue and may cause considerable damage to books. The fiye pestilential species are probably all introduced, the most common being *Ctenolepisma longicaudata* and *Ct. urbana* which attain a length of 15mm.

At least one Australian family, Ateluridae, has been found living with termites and ants (in southern Australia, the Northern Territory and in Western Australia). Their size (3–6mm in length) renders them inconspicuous. Others live under tree bark or in caves. All the native species are said to be vegetarian.

When mating, the male deposits his sperm case *(spermatophore)* on the ground, to be retrieved by the female. The female then puts the sperm into her sperm store *(sper-*

matheca). Some females lay eggs only when they moult but others lay single eggs throughout their lives. Eggs take 10–60 days to hatch and sexual maturity is usually reached about 3 months after hatching.

Despite their small size, the silverfish have an average life of 4 years and reproduction occurs throughout this period. Silverfish pass through a continuous series of moults which occur throughout their lives. Families are identified by the presence or absence of eyes, by body shape and length and by tail appendages.

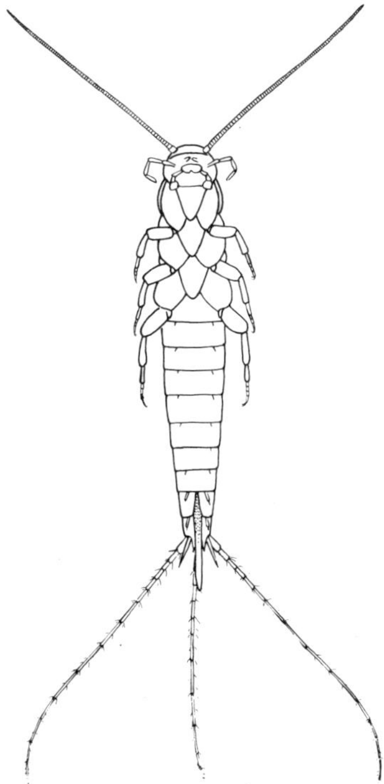

Silverfish, ventral view.

Primitive-Winged Insects

Infraclass PALAEOPTERA (old wings)

Considered to have evolved earlier than other living forms of winged insects, this infraclass contains only two orders—mayflies, and dragonflies and damselflies. None of these insects is able to fold its wings back against its body when at rest, although each holds its wings in a different position.

These insects are all closely associated with freshwater, and in both orders the life-cycle is an incomplete metamorphosis. The free-flying adults mate, but the female always lays her eggs on plants alongside water, or directly into the water. These hatch into wholly aquatic nymphs, found in many different areas.

It is considered that while both these orders are primitive, each probably originated from widely differing ancestors which in turn must have developed independently at an early stage in their evolution.

MAYFLIES

Order EPHEMEROPTERA (short-lived with wings)

As the name of the order suggests, these insects have a short adult life, and are unable to eat once they have emerged from the final nymphal moult. Of the two pairs of wings, the rear are much smaller than the front. Mayflies are easily recognised at rest, as they hold their veined transparent wings erect and have two or three long drooping 'tails' *(cerci)* which extend noticeably behind the abdomen. Beneath them are two much shorter 'claspers' *(gonostyles)* used to hold the abdomen of the other insect when mating.

In Australia, 124 species of mayflies have been recorded and each has a distinct habitat. These range from lowland bogs and shallow pools to sub-alpine streams and large lakes and rivers.

Note: For mayfly nymphs, see 'Aquatic Insects' (pp. 214–5).

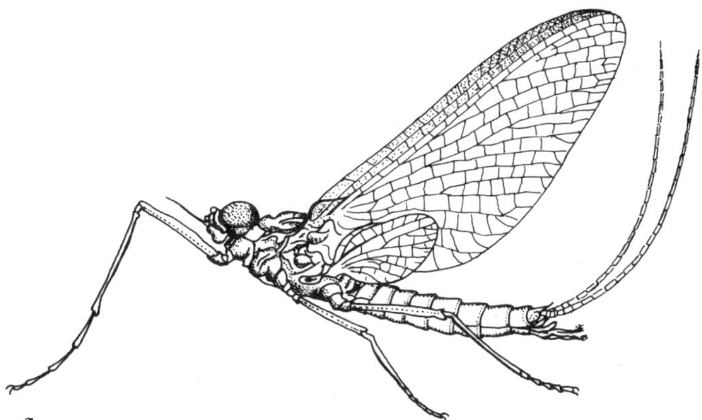

Mayfly.

DAMSELFLIES AND DRAGONFLIES
Order ODONATA (toothed insects)

Insects in this order are easily recognised by their long, straight abdomens, by two pairs of near-equal length, long wings, and by their large compound eyes. Their distinctive appearance, their brilliant colours (reds, blues, greens and gold) and their swooping, darting, hovering flight near ponds and lagoons, are all aids to identification.

Damselflies (Suborder Zygoptera) are more delicate than dragonflies, have a rather slower flight and, when at rest, hold their wings erect over the body, pressed back-to-back. The front and back wings are similar in shape and vein pattern. The eyes are smaller, and set further apart than the dragonfly's.

Dragonflies (Suborder Anisoptera) hold their wings horizontally when at rest. Front wings are of a different shape and venation from the back wings, and the eyes are exceptionally large and set close together, almost touching in the middle.

There are 97 described species of damselfly in Australia and 166 species of dragonfly. Most are between 30 and 90mm long but the largest dragonflies reach a length of 150mm, while the smallest never exceed 20mm.

Fossil dragonflies are known from the Permian era (270 million years ago) and were giant compared with contemporary species. The largest, whose fossil impression is now housed in the Paris Museum, had a wing span of 700mm.

Dragonflies are one of the fastest flying insects and have been timed at a speed of 14m per second.

All adult Odonata are carnivorous and attack and capture other insects in the air. Their prey includes mosquitoes, bees, robber flies, butterflies and other large insects.

In some places, dragonflies have been called 'bee butchers' and 'mosquito hawks', names which aptly describe the voracity and fierceness of their predation on other insects. However the earlier names of 'devil's darning needle', 'snake feeder' and 'horse stinger' were inapt, for dragonflies are not known to attack mammals or reptiles. Despite their speed and strength, they also have their enemies and adults often become enmeshed in spiders' webs or are eaten by insectivorous birds.

The means of male sperm transference in the Odonata is unique among insects (Williams 1968) and is rather similar to that of the spider, which produces sperm in its abdomen and has to transfer it to the female by means of its palp. The male Odonata has a true genital aperture where sperm is produced on its ninth abdominal segment, but its penis and sperm storage sac are on the second and third abdominal

Left: Example of male damselfly *Synlestes weyersi*. Wingspan 70 mm.
Right: Example of male dragonfly *Aeschna brevistyla*. Wingspan 80 mm.

segments. Before mating, the male must transfer sperm from the genital aperture to its penis. Dragonflies and damselflies perform darting courtship displays which are often seen by pond or lagoon in the summer months. Sometimes they fly paired, but without genital contact. Mating, when it occurs, may be in mid-air or on nearby vegetation.

After mating, some species drop their eggs directly into the water, like a mayfly, while others, with a specially-adapted ovipositor, make slits in plants growing in or near the water and into them insert elongated eggs.

Note: For damselfly and dragonfly nymphs, see 'Aquatic Insects', (pp. 216–7).

Advanced-Winged Insects

Infraclass NEOPTERA (new wings)

In all remaining orders of insects, the wings may be folded at rest along the body. Some specialised forms such as fleas (Siphonaptera) have had the wings eliminated in response to particular adaptive needs.

BLATTOID—ORTHOPTEROID ORDERS

Cockroaches	BLATTODEA
Termites	ISOPTERA
Praying mantids	MANTODEA
Earwigs	DERMAPTERA
Stoneflies	PLECOPTERA
Grasshoppers and Crickets	ORTHOPTERA
Stick insects	PHASMATODEA
Web-spinners	EMBIOPTERA

All insects in these orders share the following features:

* Mouths have distinct toothed jaws (mandibles) adapted for biting and chewing.
* Forewing is usually thickened *(tegmen)*
* Hind-wing usually has a large, fan-shaped lobe at the back near the body
* Most have Malpighian tubes

* Larvae develop wings outside the body as wing buds *(exopterygote)*
* Metamorphosis is gradual and slight, most embryos developing progressively through a series of moults *(instars)* but with no resting stage. The nymph differs from the adult only in its lack of wings and reproductive organs.

GUIDE TO IDENTIFICATION

Body flattened and wide, wing veins primitive, legs adapted for running, eggs enclosed in a hardened egg case *(ootheca)* . . .**Cockroaches** (pp. 10–13)

Bodies soft and sausage-like, workers and soldiers blind, soldiers with large heads, some with wings with primitive venation . . .**Termites** (pp. 13–20)

Long narrow body with lengthened thorax and triangular head, slender middle and hind legs, toothed front legs, eggs contained in an ootheca. . .**Praying mantids** (pp. 20–4)

Long slender body, leathery forewings short and strong, hind-wings large, transparent and semi-circular, forceps on end of abdomen, short legs . . .**Earwigs** (pp. 25–6)

Membraneous wings, hind-wings wider than forewings, three segments on tip of leg, long thread-like antennae, small head . . .**Stoneflies** (pp. 27–8)

Body long, hind legs much longer and stronger than others and specialised for jumping, wing rudiments reverse direction in later stages . . .**Grasshoppers and crickets** (pp. 28–38)

Long slender body, long spindly legs, eggs free and thick shelled, winged rudiments of nymphs do not reverse direction in later stages . . .**Stick insects** (pp. 39–42)

Short elongated body, largish head, noticeable jaws, wings found only on males, simple venation, enlarged lower front legs . . .**Web-spinners** (pp. 42–3)

COCKROACHES
Order BLATTODEA

Despite the revulsion that cockroaches often inspire among people whose homes they invade, from the entomological viewpoint these insects are one of the finest examples of adaptation to the changing environments of this world.

One of the earliest known fossils of a winged insect is the imprint of a pair of jaws *(mandibles)*, possibly of an ancestral cockroach, found in middle-Devonian sandstones of Scotland, some 400 million years old. Even if this fossil is discounted, large numbers of primitive cockroach fossils

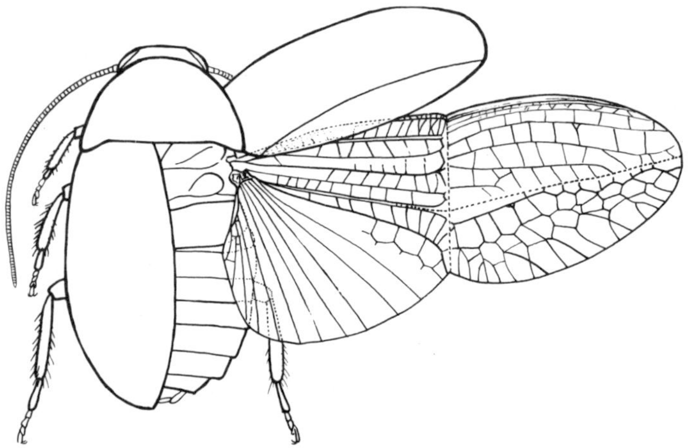

The rare *Diploptera dytiscoides* (Northern Australia) with complex folding wing. Length 14 mm.

are known from the Carboniferous, the period following the Devonian and ending 350 million years ago.

Like cockroaches today, these ancient insects had a broad flattened oval body, the front of the thorax often hiding the head which has thread-like *(filiform)* antennae of varying lengths. The wings lie flat on the back and the narrow and thickened forewings *(tegmina)* are specialised protective structures. The hind-wings are broadened at the rear by the formation of an anal fan *(neala)*. This is one of the characteristics now used to determine the difference between some of the five families present in Australia.

Some species are well adapted to flying but others lose their wings soon after passing into maturity. Some Australian

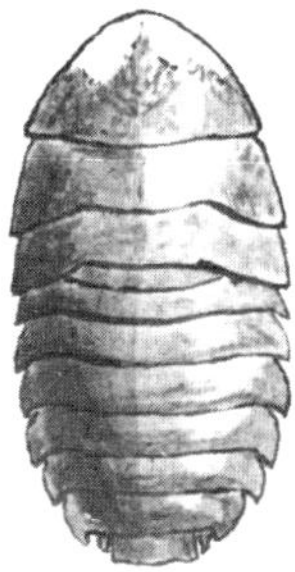

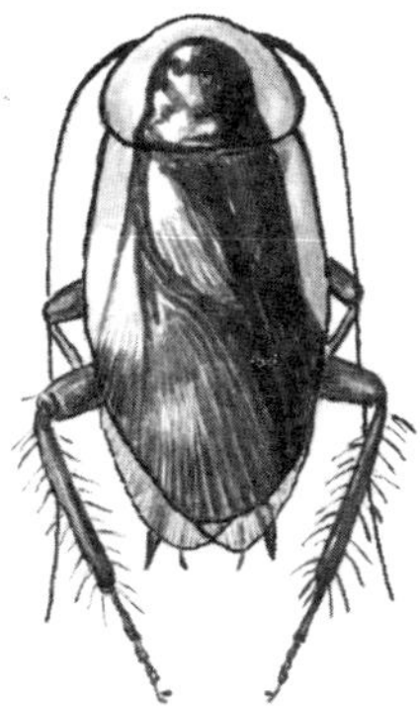

Left: Wingless female *Calolampra irrorata*. Length 30 mm.
Right: Example from Blattidae family; male *Methana marginalis*. Length 35 mm.

species are wingless *(apterous)* throughout their life. The legs of all cockroaches are well adapted to running *(cursorial)* and their flattened bodies enable them to retreat into cracks in the bark of trees or between boards of a house.

After simple mating, reproduction can occur in one of three ways. Some species develop a hardened egg case *(ootheca)* containing 12 to 40 eggs which they may carry for varying periods. Others are *ovoviviparous*, the young incubating within a brood sac inside the female. One overseas species is viviparous.

Metamorphosis is incomplete and the growing young may take from two months to more than a year to pass from nymphs to mature adults. Smaller cockroaches live shorter lives than the larger species and the adult's lifespan can vary from one to four years. Few cockroaches however, live their full lifespan. Eggs are parasitised by other insects, and nymphs and adults are attacked and eaten by other insects and by frogs, reptiles and insect-eating mammals including bandicoots.

Cockroaches possess scent glands. Some are used for defensive purposes and others to attract females. In the wild, Australian native cockroaches are adapted to many habitats ranging from the arid interior to tropical rain forests, and from sea-level to altitudes above 1300 metres. While most species are nocturnal and adapted to living in narrow cracks beneath bark, under stones and like places, others are diurnal and bask in the sun. One species is completely blind *(Trogloblattella nullarborensis)* and is found in caverns beneath the Nullarbor Plain.

The majority of cockroaches feed on rotten wood or other vegetable matter but some species are said to be omnivorous, feeding also on carrion. Domestic species can become pests, eating scraps of household food and even occupying cupboards and eating labels off jars and cans.

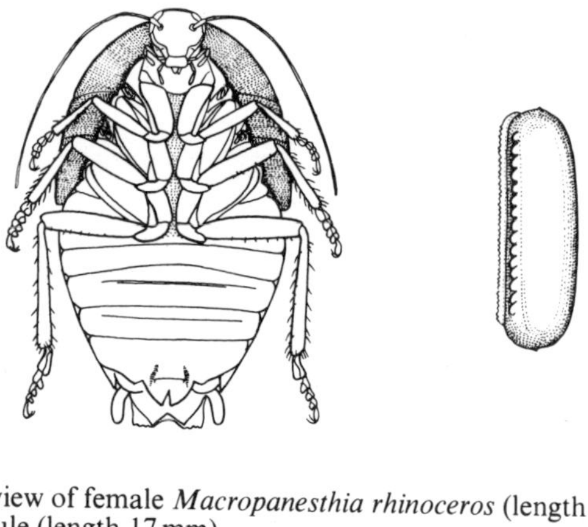

Ventral view of female *Macropanesthia rhinoceros* (length 45 mm) and egg capsule (length 17 mm).

In Australia, 439 species of cockroaches have so far been identified and are classified into four families. Five of the most pestilential species found in east-coastal homes are included in the Blattidae. They are identified by having spines on the underneath or back surface of the tarsal segments of the front legs and on the second segment of the mid and hindleg tarsi. These species include the Oriental cockroach *(Blatta orientalis)* and the American cockroach *(Periplaneta americana)*.

All except one species of Australian cockroach in the Blattidae family are flightless. Some still have traces of, or reduced tegmina, while others have lost all traces of wings. Colours of the Australian species may range from black and brown to red and even iridescent green in *Polyzosteria viridissima*. Species in the genus *Cosmozosteria* may have distinctive yellow bands or spots on a reddish-brown background. Others have pale borders or stripes. The pestilential German cockroach *(Blattella germanica)* is from the family Blattellidae.

In the family Blaberidae, the species in the genus *Panesthia* are often found in rotting logs. Their wings are broken off soon after maturity. Australia's largest native cockroaches, in the genera *Macropanesthia* and *Geoscapheus*, are wingless and burrow in soil. They are found in the coastal rain forests of north-eastern Queensland with the species *Macropanesthia* growing to a length of 75mm.

TERMITES

Order ISOPTERA (equal-wings) castes in genus Coptotermes

Termite mounds conspicuously dot the landscape throughout the outback of Australia. Some are less than a metre high; others, in warmer areas, may achieve a height of 6m or more and may have a diameter as great as their height. But the mounds built by some earth-dwelling termites only represent a part of Australia's rich termite population. The others, if soil-dwellers, may build nests completely underground. Yet another group build their nests within wood and their activities may remain unknown until the damage they cause is revealed.

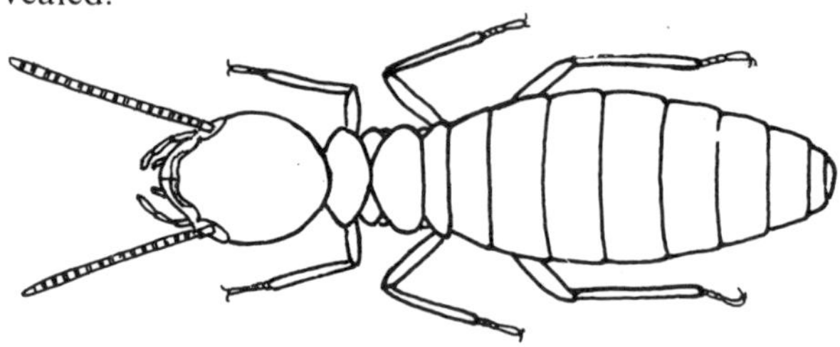

Hamitermes silvestrii worker, Central Australia. Length 4 mm.

The name 'white ant' is erroneous, the only similarity or relationship between termites and ants being that both are social insects, living in large communities. Termites undergo no metamorphosis and nymphs are minute replicas of the adults.

Modern entomologists consider that the termite's closest relative is the cockroach—this relationship being seen most clearly in the primitive soil-dwelling Australian giant termite *(Mastotermes darwiniensis)* which has five-segmented feet *(tarsi)*, an anal wing lobe and numerous Malpighian tubes, all in common with the cockroach.

Otherwise, termites resemble neither cockroaches nor ants. The termites have long, sausage-shaped, yellowish bodies, with a large head in the case of soldiers and workers.

Every termite colony is divided into castes, though the number of these social orders in any colony may depend on the species. The basic castes are the winged reproductive members *(alates)*; the wingless, blind and sterile workers; the wingless, sterile and blind soldiers with exceptionally large heads and frequently heavy jaws; and the potentially reproductive termites which are maintained to develop and expand the population.

Modern identification of families is on two bases. Alates

are identified by the number of segments on the tarsi, the number of segments on the antennae and by their wing lobes. Soldiers are identified by their tarsi, their cerci and the structure of their powerful heads.

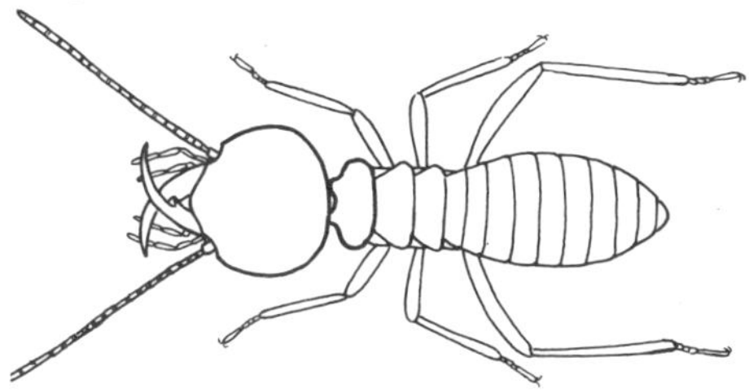

Mandibulate soldier of *Hamitermes silvestrii.* Length 4.5 mm.

On a species level, soldiers may also be identified by the shape of the strong chewing jaw *(mandible)*. Most termite soldiers have mandibles but some, in the genus *Nasutitermes*, have flask-shaped heads, the front being elongated and forming a snout *(nasus)* which opens in the front. Through the nasus, they can eject a chemical which will repel or delay intruders in the nest.

Colonies are founded when large numbers of winged alates swarm and depart on colonising flights. Soon after alighting, the alates shed their wings. Females (queens) attract the males (future kings) by releasing an attracting scent. The pair then seek a suitable nesting site in earth or wood and begin to excavate a nesting chamber. Then mating occurs. In some species with extensive nests and galleries, new colonies may be started in other parts of the general nest structure. The queen soon begins to lay her eggs and these hatch some weeks later. The larvae are tended by the king and queen until they are large enough to take their place in the new society.

It seems that a newly-hatched nymph may develop into any one of the termite castes. In some cases, it is probable that the caste is determined by the kind of hormone the parent transmits to the nymph by licking, or the kind and quality of food the parent supplies.

The king and queen do not produce alates for some years, and the colony is at first maintained solely by the royal parents and by workers and soldiers. During this time, the number of workers increases and the queen lays increasing numbers of eggs. Some queens eventually produce between 2000 and 3000 eggs, causing their abdomens to become considerably distorted. In time their abdomens become immobile and they have to rely entirely on workers to feed them.

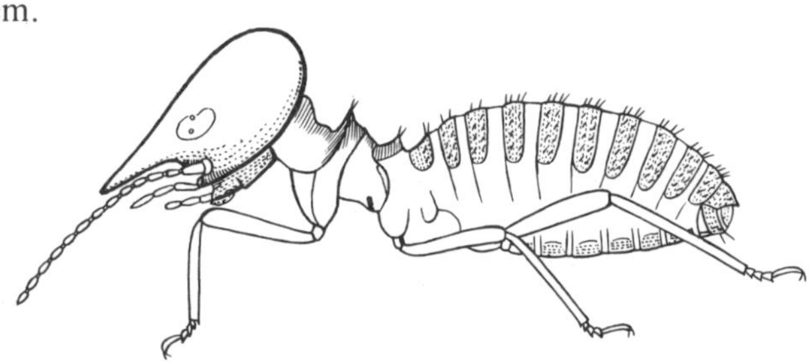

Nasute soldier from *Eutermes vernoni*. Length 3 mm.

Termites feed on dead wood, vegetable matter and fungi, both living and dead. Many termites possess a form of protozoa, a living organism, in their abdomen which converts cellulose in wood into a digestible material. Those termites without protozoa are assumed to possess some form of stomach bacteria which does the same job, or they may produce an enzyme to dissolve the cellulose.

From the basic nuptial chamber of the queen the termites build a nest. This may be in a wooden stump, or in the soil. The colony excavates a series of galleries into the wood or, as time passes, digs a series of tunnels to reach supplies of food above or below the surface. Both wood-dwelling and earth-dwelling termites feed on wood, but there is much variation in nest design and structure.

The conspicuous termite mounds often seen in the bush represent the termitarium with a highly-organised structure.

Other soil dwellers live and breed in a system of extensive subterranean galleries, some extending for a distance of 90m or more from the nursery area and covering as much as a hectare. Other nests extend several metres below the surface.

Wood-dwelling termites do not construct external nests but live entirely in galleries excavated in the wood on which they feed. They do not need to remain in contact with the ground and can live in timber isolated from the soil.

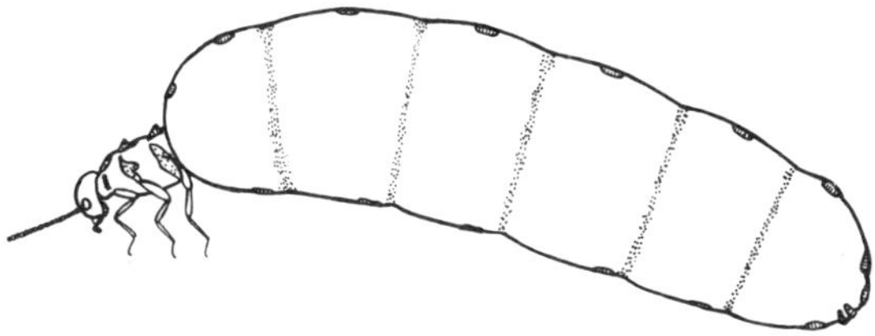

Coptotermes lacteus; fertile queen. Length 17 mm.

Some termites may eat grass and forage at nights in the open. With these exceptions, all termites avoid light and when extending their range, always build protective tunnels or tubes, often on the outside of materials they cannot penetrate. These tunnels allow the termites to travel quickly between nest and feeding areas.

Termites require a high humidity and within the nest, a relative humidity of 90 per cent is maintained. This is achieved both from the environment and by the respiration of the termites. Although some termites can live in dry wood, surface galleries of the grass-eating *Hamitermes* have been shown to be deserted in times of drought, the colony retreating below ground-level to deeper chambers where the humidity is higher. Paradoxically, although termites need high humidity within the nest, they are more commonly found in arid areas than in rain forests.

The age of termite mounds has been investigated. One mound of *Nasutitermes triodiae* was estimated to be 150 years old while another, near Darwin, has been in existence for at least 30 years.

With some species, the mound does not achieve maturity until it is 25 years old and such colonies have a natural life of at least 50 years. Grass-eating termites including *Hamitermes* species have a remarkable capacity for fast mound building. In the outback, new landing strips for aircraft are cleared of the inevitable termite mounds. Unless the soil beneath these mounds is poisoned, after the first shower of rain a new mound is rapidly built by the subterranean colony and, in a short time, these structures can become large enough to endanger aircraft using the strip. Other species can speedily repair ravages to their mounds and *Nasutitermes exitiosus*, for example, can survive raids by echidnas and withstand repeated attacks by ants.

A mature termite nest may have a population of hundreds of thousands—sometimes more than a million inhabitants; however, a whole colony may be killed by meat ants *(Iridomyrmex purpureus)* who can completely occupy a termite nest and kill all the former inhabitants in a surprisingly short period.

If opened, most termite mounds may be seen to consist of an outer casing of clay several inches thick and extremely hard. Within this casing, there is a structure of 'paper' or digested wood, a honeycomb of passages and galleries all centred round the royal cell which is shaped like an inverted saucer. The nursery, where eggs are placed, is above the royal cell. In some overseas species, termites cultivate fungus which they feed to the nymphs. No Australian species has been discovered doing this, although decaying wood is often favoured by termites as a nest location. Termites are cannibals and may eat their fellows to supplement their diet and to prevent the colony growing too large at times when nesting flights cannot be made. Other termites seal away their dead in separate chambers and also have other chambers which serve as waste disposal areas, being filled with excrement and other rubbish, before being sealed.

Nest structures of termites are often distinctive and large. The world-famous magnetic termites *(Amitermes meridionalis)* of northern Australia, build tall narrow nests,

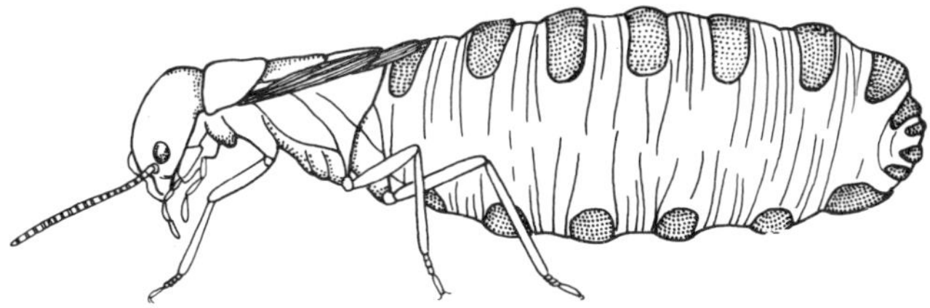

Hamitermes silvestrii; neoteinic, or emergency replacement, queen. Length 12 mm.

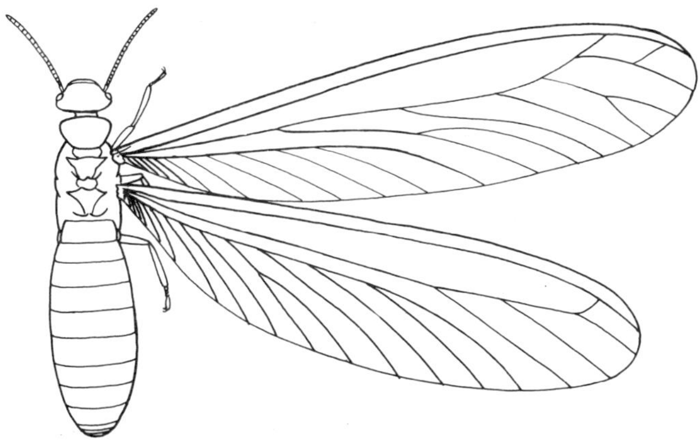

Hamitermes winged male. Forewing length 15 mm.

often two metres high, which always have the points facing north and south. *Nasutitermes triodiae* build tall structures like columns which can be more than 6m high.

Not all distinctive termite nests are on the ground. Two species of *Microtermes* build clay covered nests on poles or posts; and two species of *Nasutitermes* build bulbous nests high in trees.

Large termite nests provide accommodation for many other forms of life. In one mound built by *Coptotermes acinaciformis* seven other species of termites were found.

Other insects also share parts of the termitarium and parrots, kingfishers and goannas may dig their nests inside termite mounds. The golden shouldered parrot of Cape York, Northern Queensland, nests only in termite mounds.

The soil and wood processed by termites has exceptional strength and for many years termite nests have been broken up, crushed, mixed with water, and used by people living in the outback to make a form of cement. Tennis courts, cricket pitches and cow yards have been made from this mixture which, when baked by the hot sun, has exceptional strength and resistance to erosion by wind and rain.

Some of Australia's termites are of primitive origin and the family *Mastotermes* is the oldest surviving group of this type. Today there are at least 246 species of termites in Australia representing five of the six families and more than 2000 species found throughout the world.

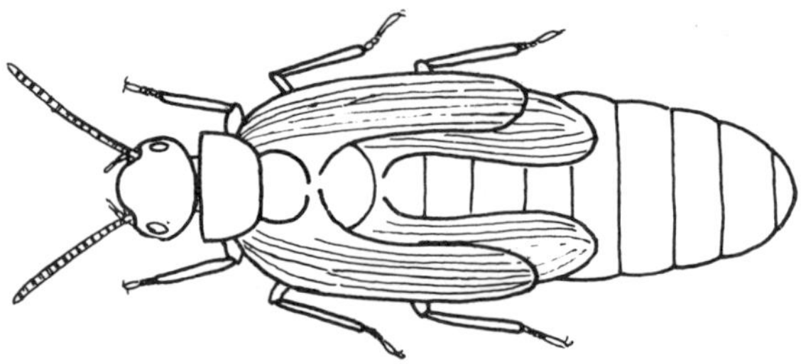

Hamitermes, a late nymph.

PRAYING MANTIDS

Order MANTODEA

With their large triangular heads, which swivel on a mobile neck, their long body, slender long hind-legs and predatory *(raptorial)* front legs, mantids are easily-identified insects found both in cities and in the bush. They fly by day and by night, and are a familiar sight in houses and gardens. The appearance and stalking posture of mantids make them quite remarkable, and although their green or brown colour allows them plenty of camouflage when perched on plants, they may frequently be seen against more contrasting backgrounds.

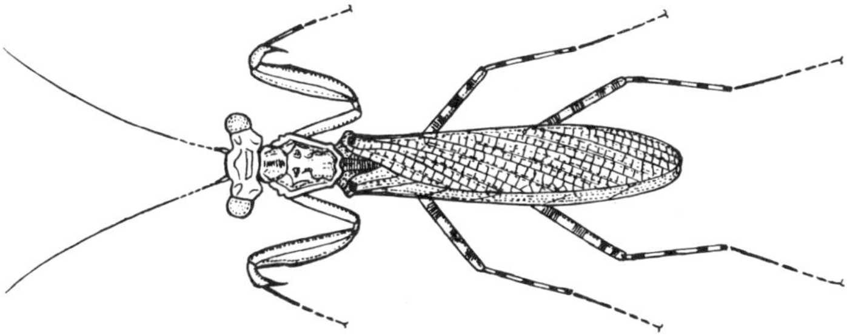

Phthersigena conspersa; an example of the small, bark-haunting mantids. Length 22 mm.

Mantids are carnivorous and their life is spent hunting other insects, although in captivity they have taken apple, potato and other raw vegetables and fruits. The large triangular head is dominated by two huge compound eyes. When poised, stalking its prey, the head turns on the slender neck, giving the insect a distinctive appearance.

The two Australian families are classified by the presence or absence of spines on part of the tibia of the forelegs. The forelegs are armed with a sharp hook and a series of spines which are brought together to pierce anything grasped in a pincer-like grip. Wings may be present on both sexes of mantids but the males' are usually functional on all species while the females' are often reduced or absent. The forewings are long and narrow while the hind-wings are almost as long but, when outstretched, are wider and more triangular.

A mantid's heavy body might suggest that its flying ability is limited but in the USA mantids have been seen on the observation platform of the Empire State Building in New York hundreds of metres above street level.

In their native habitat, mantids live solitary existences, mainly inhabiting tree trunks, bushes and tall plants. When seen, they are generally at rest, the forelegs raised in a 'praying' attitude and the body projecting upwards. If disturbed, they may run quickly, jump or fly away. Other species, especially those with brightly coloured wings, will face the 'enemy', raising their wings and striking with their spiny forelegs.

At mating time, the male mounts the female and in some species, the female may attack and eat the male during or after mating has taken place.

Like the cockroach, the female produces an egg case *(ootheca)* in which are chambers housing 10 to 400 eggs. The ootheca is made from a frothy material which the ovipositor valves mould into a distinctive shape which hardens in the air to form a tough, sculptured capsule. The ootheca is always attached to a surface—a tree trunk, a wall or even a blade of grass. In Melbourne, egg cases have been found on clothes drying in the garden and even on a clothes peg. While some females remain near the eggs until they hatch, others fly away as soon as the eggs are laid.

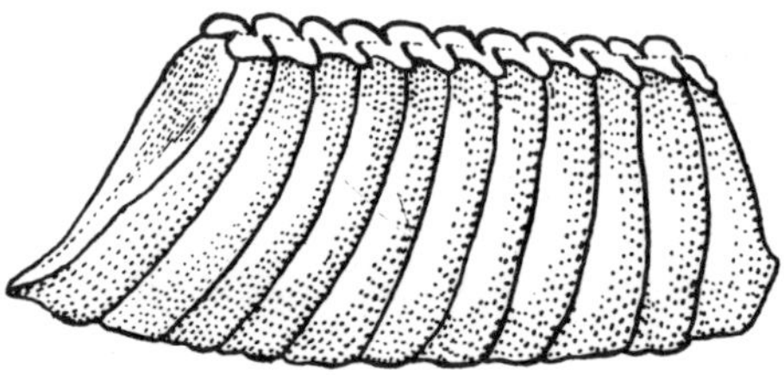

Egg capsule of *Orthodera ministralis*. Length 12 mm.

Various wasps lay their eggs in the mantid egg case, and some flies have been known to breed in them, but it is not known whether the latter are parasites or scavengers. The hatched mantids are also attacked by birds, lizards and insectivorous marsupials while wasps, mites and nematodes also take their toll.

Eggs which survive hatch into nymphs which are little different from the adult mantid. The nymph passes through several moults and during this period, wings and genital organs increase in size each time the insect emerges from its former casing.

Australian mantids are grouped into two families, the Amorphoscelidae and the Mantidae. The former have a mottled colour and a body shape with a noticeable protuberance at the rear of the abdomen. Highly specialised mantids in the subfamily Paraoxypilinae also possess spines on the head and on their short neck which gives them perfect camouflage against the lichen on which they live.

The Mantidae includes 100 Australian species compared with only 30 Australian species in the Amorphoscelidae; and these species represent only a fragment of the world's 1800 species of mantids. The subfamilies of the Mantidae are classified by the number of spines on the femur of the forelegs and on the shape and length of the neck or pronotum.

The most common Australian species is the green mantid *(Orthodera ministralis)* which grows to 30 to 50mm long and has a wide neck. The brown mantid *(Archimantis latistyla)* found in both coastal and inland areas of eastern Australia, grows to a length of 100mm and is pinkish-brown. Near Sydney, *Tenodera australasiae* is common and is recognised by its long slender wing covers *(tegmina)* which are bordered with pink and green. The mottled hind-wings have a pink leading edge.

In America, mantids have long been kept as pets and for classroom study. A suitable cage may be made from a fish tank with mosquito netting above, or wholly from fine rigid mesh. Mantids make an interesting subject for study though due to their cannibalistic tendencies, only one mantid should be kept in a cage at a time. Mantids may be fed live or dead insects, crumbs of meat and even fruit and vegetables. In time, it is said that they will take food from the fingers and will eagerly sip water from a spoon. Mantids have been conditioned to expect food from their keepers and are said to recognise the person who feeds them.

Eggs laid in captivity may also be kept and incubated. In the southern states, they are laid in autumn (often in May) and may take from four to six months to hatch. The egg case should be housed in a separate cage (a jam jar with netting on top serves well) so that it will not be parasitised by wasps. Also, when the young hatch they may be eaten by the mother, and cannibalism may also occur among the young.

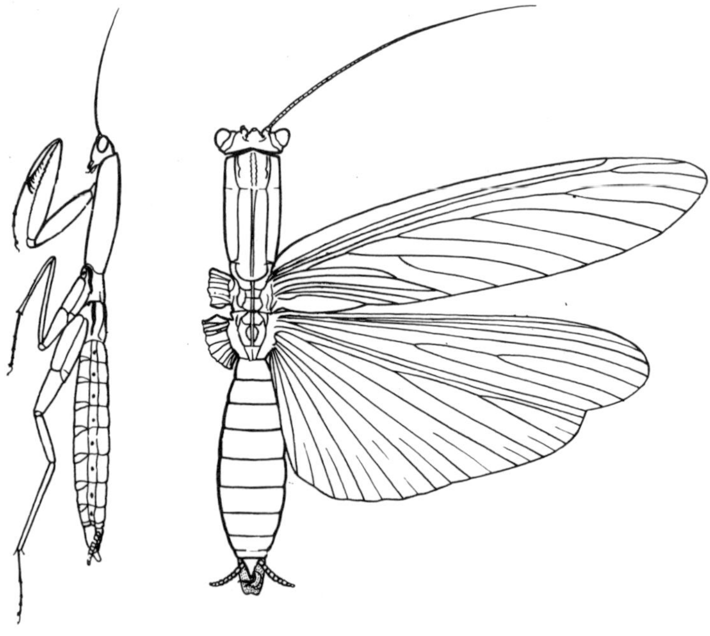

Common green mantid; *Orthodera ministralis.*

EARWIGS
Order DERMAPTERA (skin-winged)

Long-bodied, slender, usually brown with a tough integument, earwigs are easily identified by the long pair of pincers (modified or specialised cerci) at the back of the abdomen. Australian species rarely exceed 25mm in length and their legs are short, and like the cockroach's, are adapted to move quickly and enable the earwig to scuttle into narrow spaces.

Not all earwigs have wings but when present, the thick forewings *(tegmina)* are short and strong, like those of a beetle. The rear wings are large, transparent and semi-circular. When not in use they can be folded so that they are almost hidden beneath the forewings.

Earwigs are nocturnal, but may be attracted by lights and are not uncommon visitors to houses and other domestic buildings. They appear to be omnivorous, killing other insects but also eating plants and carrion. The pincers seem to be used as 'hands' to carry prey after it has been killed, or to hold the mate when copulating. Some earwig populations have a deficiency of males. As a consequence, females are able to retain the male's sperm for long periods after mating.

Unlike most insects, the female earwig is renowned for the care she lavishes on her eggs and young. After laying 20 to 50 white eggs in the tunnel she digs for this purpose, she remains with them until they hatch, constantly turning them and keeping them clean, and keeping predators at bay. Hatching occurs in two or three weeks and the nymphs are tiny replicas of the adult. Before maturity is reached, nymphs may moult four or five times.

Except for damage caused to some fruit, vegetables and grain in Queensland and New South Wales by two imported species, earwigs seem harmless to man. One species, *Labidura truncata* is said to attack codling moth larvae.

Some 60 species of earwigs are known in Australia, a

small fraction of the 1200 species identified throughout the world.

Australia's largest earwig, *Titanolabis colossea*, attains a length of 35mm and is predominantly black in colour with relatively short, blunt and heavy pincers. In contrast, *Apachyus australiae*, found in coastal forests of New South Wales and Victoria, is beautifully coloured with markings in reddish brown, cream and black.

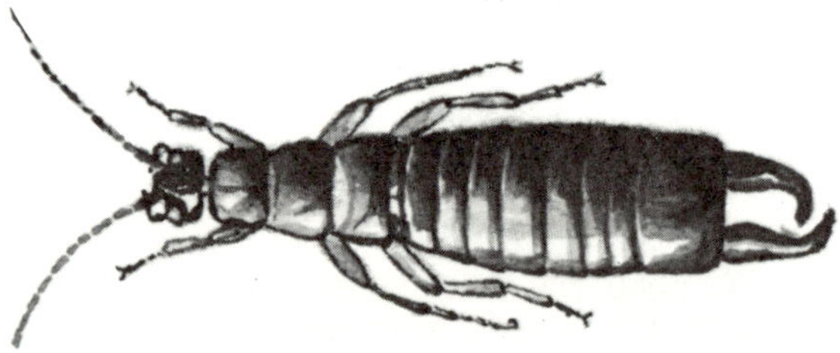

Anisolabis littorea male showing characteristic pincer development in this species. Length 30 mm.

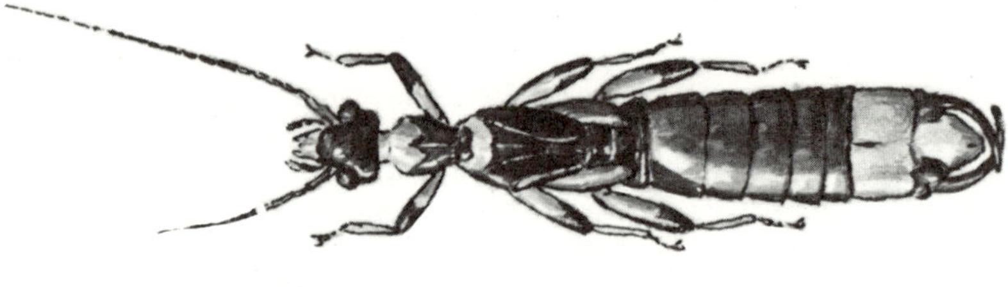

Apachyus australasiae, male. Length 40 mm.

STONEFLIES
Order PLECOPTERA (folded winged)

Always found near water, stoneflies vary from less than 5mm to 50mm in length. The gauze-like wings are long, the front pair are narrow but the hind pair are of much greater width, and may be folded like fans. Stoneflies have long, thread-like *(filiform)* antennae and two cerci, sometimes as long as the antennae.

Overseas, stoneflies are sometimes known as 'salmon flies'. Their aquatic nymphs (see pp. 218–9) are an important source of food for fish, especially salmon and trout.

Stoneflies are usually found near fast-flowing water or occasionally on the edges of unsheltered lakes. They are

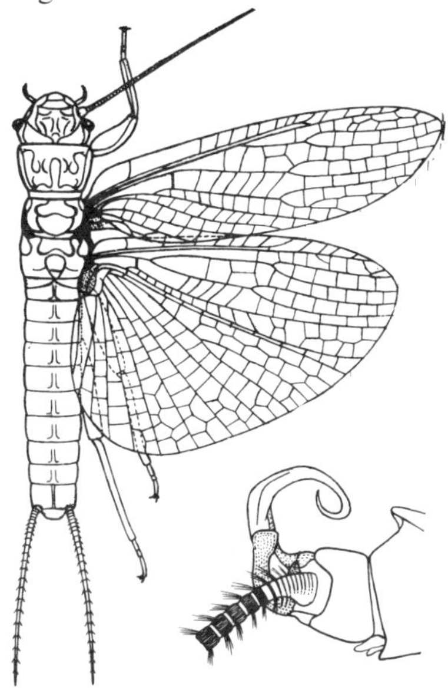

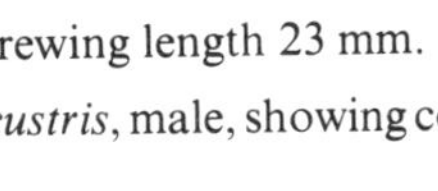

Left: *Eusthenia spectabilis*, female. Forewing length 23 mm.

Right: End of abdomen of *Eusthenia lacustris*, male, showing copulatory mechanism.

rarely found near stagnant or slow-running water. Stoneflies occur from Tasmania to Queensland, including the Kosciusko region and other places of high altitude. Only two species have been recorded in Western Australia.

Mating occurs when the stoneflies are at rest and subsequently the female drops her eggs into the water during her slow and laboured flight, the tip of her abdomen penetrating the surface as the eggs are dropped. This habit is not universal however and some species let their eggs fall freely from above the water or from an overhanging rock or log where the female is resting.

Possibly because fish and other aquatic fauna live off stonefly nymphs, very large numbers of eggs are laid. One Australian species is estimated to lay at least 1000 eggs at a time. Although more than 1000 species in eight families of stoneflies have been identified, Australia has only 84 species. Identification of adults is based on wing venation and of nymphs on the arrangement of gill structures.

During the day stoneflies may be found resting on tree trunks alongside water. With their dark wings folded flat over their backs and well camouflaged against the bark, they are easily passed by.

Most Australian stoneflies are dull in colour but those in the genus *Eusthenia*, found only in Tasmania, have five species whose wing under-surfaces are a bright red or purple.

Note: For stonefly nymphs, see 'Aquatic Insects' (pp. 218–9).

GRASSHOPPERS, LOCUSTS AND CRICKETS

Order ORTHOPTERA (straight-winged)

The name 'orthoptera' refers to the way in which insects in this order fold their wings lengthways along the back. They also have hind-legs adapted for jumping *(saltatorial)* with long, broad muscular thighs *(femurs)* and straight rigid shanks *(tibia)*. The forewings *(tegmina)* are leathery and the hind-wings are thinner and transparent, on some species extending to the tip of the abdomen or beyond when folded. In other species, part of the body may remain exposed and a few species are wingless. Although many insects make sounds by rubbing parts of the body together *(stridulatory organs)*, with the exception of cicadas (Hemiptera, see pp. 54–5) the 'singing' of Orthoptera is the best known and loudest to the human ear.

Stridulatory organs are composed of two adjoining parts of the body, mutually adapted to making sounds by rubbing. The fixed member, which may be likened to the strings of a guitar, has small transverse toothed ridges and is known as the *pars stridens* or 'file'. The other member, the *plectrum*, is roughened, and rubs on the pars stridens like a bow on a violin. There may also be an area of tightly stretched membrane, which acts as an amplifier. These organs produce a rasping noise. When made in flight, this may be to foil predators, but usually it is part of mating behaviour, and may in some cases be the sole province of the male. Special 'ears' to receive these sounds are also present.

There are more than 20 000 species in the Orthoptera and the 2000 species in Australia represent all groups in the order. The groups are:

Suborder	*Common name*	*Superfamily*
ENSIFERA	True crickets (including tree and mole crickets)	GRYLLOIDEA
	King and cave crickets	GRYLLA-CRIDOIDEA
	Long-horned grasshoppers	TETTI-GONIOIDEA
CAELIFERA	Dun-coloured grasshoppers	TETRIGOIDEA
	Pigmy mole crickets	TRIDACTY-LOIDEA
	Short-horned grasshoppers and locusts	ACRIDOIDEA

As with other insects in these orders, Orthoptera nymphs are miniatures of their parents and differ only in their lack of wings or, as they mature, in their wing rudiments, which reverse direction as later moults occur.

Suborder **Ensifera**

Predominantly nocturnal insects, the king, cave and true crickets and the long-horned grasshoppers can be distinguished by their antennae which have more than 30 segments. Also, the 'ear', when present, is located on the tibia of each foreleg, just below the knee, and the stridulatory mechanism is in the overlapping horizontal part of the folded wings.

CRICKETS (Superfamily GRYLLOIDEA)

The three families in this group are distinguished by their three-segmented tarsi (all other insects in Ensifera have four-segmented tarsi). They are: the true crickets; the small Myrmecophilidae (which live only in the nests of certain ants); and the mole crickets.

TRUE CRICKETS (Family GRYLLIDAE)

Most crickets are nocturnal and their singing is most apparent on a warm, still evening after a shower of rain. Some species retreat beneath logs, stones and debris during daylight hours. One of the most common species is the field cricket *(Gryllus servillei)* a solidly-built insect about 25mm in length, found both in the bush and in gardens of Australian towns. In daylight, these crickets remain hidden in burrows in the soil or under stones. Females have a long, slender ovipositor and eggs are laid in the soil. The larger, black field cricket *(Teleogryllus commodus)* is said to damage pastures in south-eastern Australia.

TREE CRICKETS (Subfamily OECANTHINAE)

These insects live on exposed foliage in trees and bushes. The genus *Oecanthus* includes the ghost crickets, semi-transparent arboreal insects which inhabit the foliage of

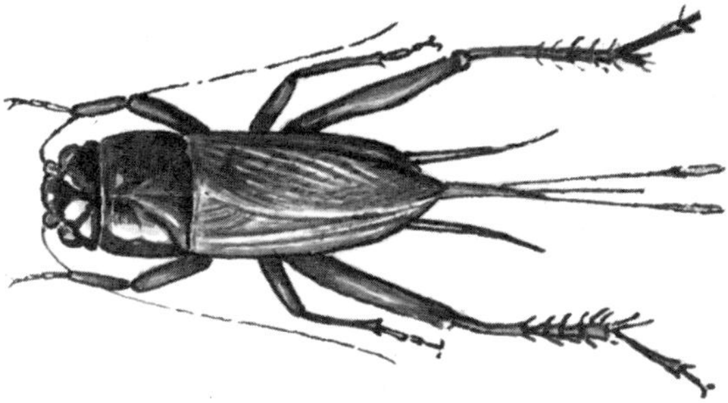

Female of *Gryllus servillei*, the common field cricket. Body length 30 mm.

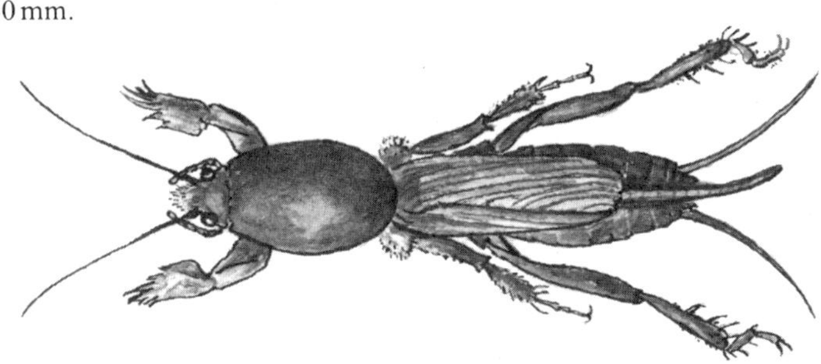

Gryllotalpa coarctata, a common mole cricket. Length 45 mm.

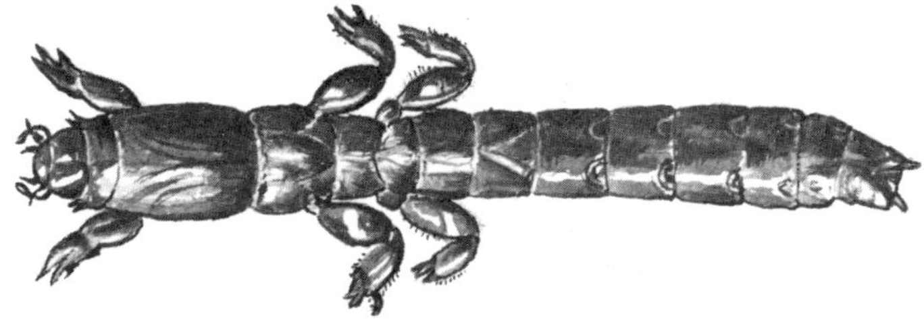

Cylindracheta kochi, plant stem burrowing mole cricket from Western Australia. Length 60 mm.

certain shrubs. The male has a cup-like depression in its thorax which contains a liquid which is attractive to females. Before mating, the female drinks this liquid and eggs are subsequently laid in slits cut in twigs of vines and bushes.

MOLE CRICKETS (Family GRYLLOTALPIDAE)

Often discovered by people digging in their gardens, there are about 50 species of mole crickets in Australia. All have large forelegs, which are flat and wide and used like a shovel for digging. There are also interlocking teeth on the tibia which cut through grass roots like shears. Mole crickets live in permanent, damp burrows and have much shorter antennae than other crickets, although each antenna still has 30 or more segments. Most mole crickets have wings and may be attracted to light. Females lay their eggs in their burrows and stay with the eggs while they incubate and protect the nymphs after they hatch. Some species are said to damage crops and the cosmopolitan yellow African mole cricket *(Gryllotalpa africana)* stores seeds underground. The related Australian mole cricket *(Gryllotalpa australis)* is common and is dark brown, with dense short hairs on its thorax.

KING and **CAVE CRICKETS** (Superfamily GRYLLACRIDOIDEA)

These primitive, cricket-like insects have four tarsi; three families occurring in Australia.

WINGLESS KING CRICKETS (Family STENOPELMATIDAE)

Having no common name, these crickets are mainly known from the genus *Australostoma.* A dark-coloured species, *A. opacum* has a body length of 35mm and is usually found under logs. Other species may grow to a length of 100mm and have been reported from rain forests in southern Queensland. They are distinguished by having antennae as long as the body, a very large head and large, curved jaws. Large species are also found in the family Gryllacrididae, which includes the winged species *Hadrogryllacris magnifica*

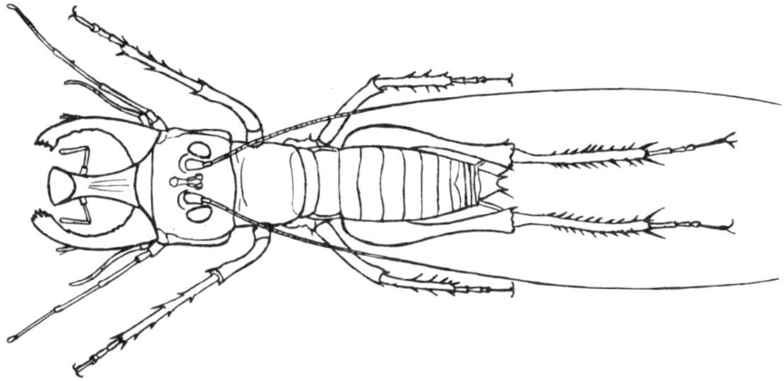

The large Australian king cricket. Length 75 mm.

found in the arid interior of Australia. Females of this species have an ovipositor or tail spine longer than their 35mm body. The head, pronotum and legs are all a shiny, iridescent black.

CAVE CRICKETS (Family RHAPHIDOPHORIDAE)

Moisture seeking insects and often found in caves, these are usually wingless and have short, soft bodies and very short antennae. Some cave-dwellers are blind and have been found in completely dark caves at Gunns Plains and Mole Creek in Tasmania.

LONG-HORNED GRASSHOPPERS (Superfamily TETTIGONIOIDEA)

Known in America as 'katydids' after the sound of their 'song', long-horned grasshoppers are mostly vegetarians though some species may predate on other insects. Eggs are laid in the foliage where adults live and while some species look like slender grasshoppers with very long antennae, others show a remarkable diversity in appearance. Australia has some 300 of the world's 5000 species of tettigonid grasshoppers and all local species are included in the family Tettigoniidae.

LONG-HORNED GRASSHOPPERS (Family TETTIGONIIDAE)

In Western Australia, *Phasmodes ranatriformis* is a long-bodied species, so slender that at first it could be mistaken for a stick insect. It is wingless and its hind-legs are not adapted for jumping.

Another long-bodied tettigoniid is *Zaprochilus australis*, which has an elongated head with a protruding jaw. When at rest, its wings rise above the body at some 30 degrees to the abdomen.

Females in *Caedicia major* have a curved ovipositor which is almost twice as long as their short, fat body. Like most of this family, the insect is coloured to merge against its natural background, usually eucalyptus leaves.

One of the most interesting species is the crested grasshopper *(Alectoria superba)*. It inhabits the arid inland of Australia and has a huge crest, larger than the head and situated just behind it.

The mountain grasshopper *(Acripeza reticulata)* shows remarkable sexual dimorphism. The slender, dark-coloured male looks like a grasshopper with elongated, fan-shaped wings. In contrast, the female is short with a fat body and has coarse-grained front wings *(tegmina)* but no flying wings. The body beneath the tegmina is black, with striking bands of bright blue and red: this seems to be a form of protective coloration for when disturbed, the female raises herself on her hind-legs, and lifts her forewings to display her bright abdomen. If further disturbed, she then erects a bright orange membraneous 'collar' behind and around the head.

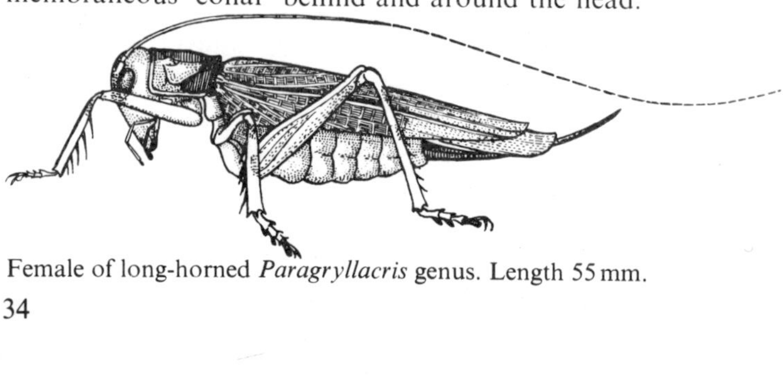

Female of long-horned *Paragryllacris* genus. Length 55 mm.

Suborder **Caelifera**

The grasshoppers and locusts in this group are mostly active in daylight and have antennae with less than 30 segments; these are generally less than half the body length. Six families are represented in Australia and contain 900 of the 10 000 species found throughout the world. The stridulatory organs of this group are contained on the forewings, which are rapidly rubbed by minute teeth on the femurs of the hind-legs. Most species are ground-dwellers, living on grass and herbs, though some may eat dead leaves when they are covered in dew.

DUN-COLOURED GRASSHOPPERS (Superfamily TETRIGOIDEA)

There are 70 Australian species of these small, dusky-coloured grasshoppers. They are usually found on bare soil or sand, sometimes on short turf, and usually in damp locations. Many species can swim and some can remain underwater for quite long periods, either resting on the bottom or clinging to water plants. They are identified by the back of the prothorax *(pronotum)*, which projects backwards and covers the first few abdominal segments. They have only three segments on each hind 'foot' or *tarsus*.

PIGMY MOLE CRICKETS (Superfamily TRIDACTYLOIDEA)

These are small insects, sometimes with shiny black heads, or with variegated colouring. They prefer a damp habitat and frequent margins of streams and lakes where they build galleries in the sand. Unlike true mole crickets (Gryllotalpidae) they spend much time on the surface, jumping to escape capture or taking to the water where, like the dun-coloured grasshoppers, they are able to swim and submerge. The pronotum of this group only just covers the thorax and there is only one segment on the hind tarsi.

SHORT-HORNED GRASSHOPPERS and **LOCUSTS** (Superfamily ACRIDOIDEA)

These are the largest and best-known group in the Caelifera, with 650 species in Australia. In the family Acrididae are found those short-horned grasshoppers which can sometimes be economically destructive.

Although used indiscriminately in the past, the terms 'grasshopper' and 'locust' have little scientific meaning from the viewpoint of identification and classification. All insects in this superfamily are short-horned grasshoppers, but some exhibit the phenomenon known as *kentromorphism*, and it is only these species (about four in Australia) which, given the right climatic conditions, are able to multiply rapidly and form heavy strongly-migrating swarms. These are known as 'locusts'.

The phenomenon of kentromorphism is not confined to locusts, but it is in these insects that its occurrence proves so costly to mankind. It also occurs with stick insects (Phasmatidae, p. 42) and some butterfly and moth larvae (Lepidoptera). Insects capable of this phenomenon possess two different physical forms or phases. The first phase is characteristic of low-density populations *(solitaria)*, the other is found only when there is heavy population pressure *(gregaria)*. Marked differences in the insect's colour and pattern, in physical size, and in physiology and behaviour occur between low-phase individuals and those in high-density populations.

A summary of Australian research into the locust problem has been given in *Rural Research in CSIRO* (No 69, March 1970). The Australian plague, or black-tipped locust *(Chortoicetes terminifera)* is distinguished by a black spot at the tip of the rear wings. Damage is also caused by the yellow-winged locust *(Gastrimargus musicus)* which produces a clicking sound in flight. The spur-throated locust *(Austracris guttulosa)* also causes damage, as does the migratory locust *(Locusta migratoria)*, a pest well-known in eastern Europe, Africa, China, the Philippines, Borneo and in the north-east canefields of Queensland.

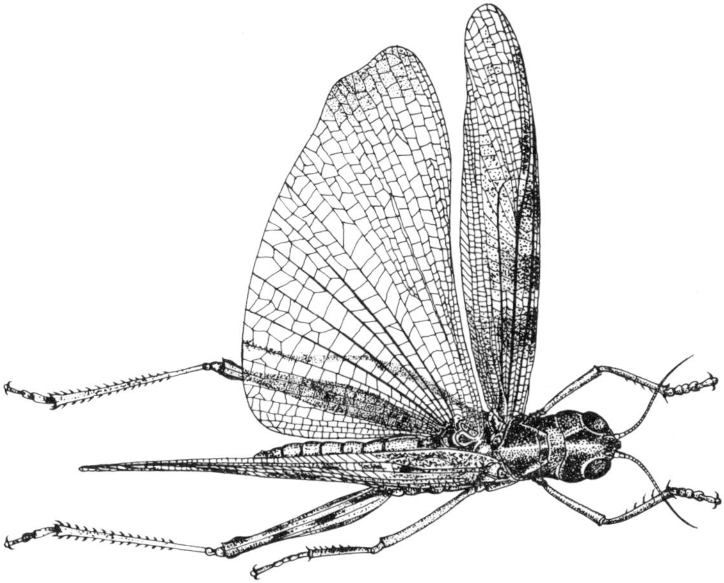

Australian plague locust.

Two other species which also cause damage are the small plague grasshopper *(Austroicetes cruciata)* and the wingless grasshopper *(Phaulacridium vittatum)*. In northern Australia, *Valanga irregularis* feeds on the leaves of trees and shrubs in the monsoonal tropics and subtropics; it is the largest short-horned grasshopper in Australia and one of the largest in the world.

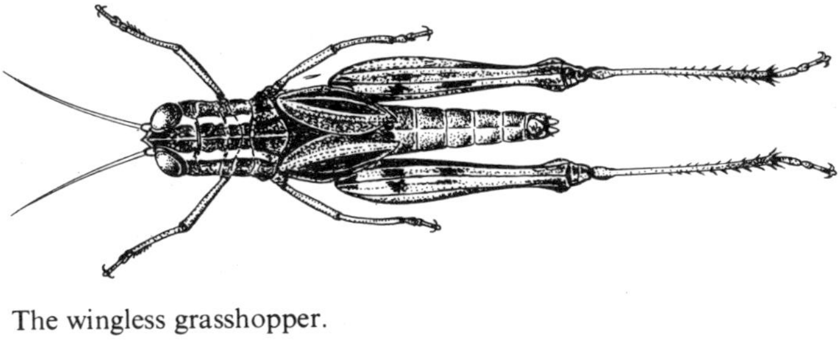

The wingless grasshopper.

The Australian plague locust normally lives in isolated and arid parts of the outback. The females have a pronged ovipositor which they can use to drill holes in hard ground by alternately opening and closing the prong and then drawing the abdomen into the hole. Into these holes they lay eggs which, if conditions are dry, may remain dormant for up to a year.

Eggs only hatch after rain, usually with the onset of cold weather. But given the right combination of circumstances, a female lays her 30 to 60 eggs and these can hatch in 17 days.

Within another 12 days, the adults mate again and a second batch of eggs hatches 17 days later. At such times, dormant eggs also hatch and greatly increase the population. Within four to six months, a plague can develop, the locusts eating both green and dry grass. They migrate in swarms by day and individually at night. Given a breeze of more than 11 km/h, the swarm will follow the wind's direction. When conditions are mild, many of the insects will seek areas where rain has recently fallen and where reproduction can continue. All grasshoppers are attacked by straw-necked ibis and by ants. There is also one wasp which parasitises locust eggs and scientists are now considering breeding these wasps to release at appropriate times in areas where breeding populations of locusts are swarming. Many eggs also die from fungus and other diseases.

Unlike locusts, plague grasshoppers breed only one new generation a year and swarms migrate only a few kilometres.

Until now, locusts have been a world-wide problem. However, research by Australian scientists into their life-cycle could lead to techniques of control which will apply wherever they are a threat to vegetation.

Among the many species of short-horned grasshoppers, two other groups should be mentioned. Some members of the Pyrgomorphidae are unusual in that they have conical-shaped heads. In the Eumastacidae, the grasshoppers are long and slender insects.

STICK AND LEAF INSECTS

Order PHASMATODEA (spectre or ghost insects)

Of all insects, possibly none are better adapted to their environment than the stick and leaf insects. The long, slender bodies and spindly legs that are finer even than those of the mantids, make the stick insect almost impossible to detect when perched on a similarly coloured twig. Also, the advanced 'leaf' camouflage of certain insects in this order provide outstanding protection from most predators.

Of the 2500 stick and leaf insects known from the warmer parts of the world, Australia has 140 species and of these, the regal phasma *(Acrophylla titan)* of North Queensland, the longest insect in the world, may attain a length of 300mm.

Unlike the mantids which they superficially resemble, stick insects are vegetarians and three Australian species cause significant damage to certain eucalypts. Their activities can kill trees by defoliation and repeated defoliation has accelerated soil erosion on steep slopes in some areas.

While most stick insects reproduce by normal mating, some species appear to suffer a deficiency of males and the phenomenon of *parthenogenesis* or 'virgin birth' has been widely studied in these insects. Parthenogenesis occurs in other types of insects but the size of the species *Carausius morosus* and the hatching of eggs from females which have never been inseminated has been widely studied in laboratories. In the Australian species, the tesselated leaf insect *(Ctenomorphodes tessulatus)*, unfertilised females not only produce eggs which hatch, but these eggs produce male as well as female offspring.

The true leaf insects of south-east Asia and New Guinea are grouped in the family Phyliidae. Their bodies are much expanded at the rear and the legs have wide horizontal extensions which contribute to their leaf-like appearance. Two species of *Phyllium* are found in the rain forests of North Queensland but otherwise, true leaf insects are not found elsewhere in Australia.

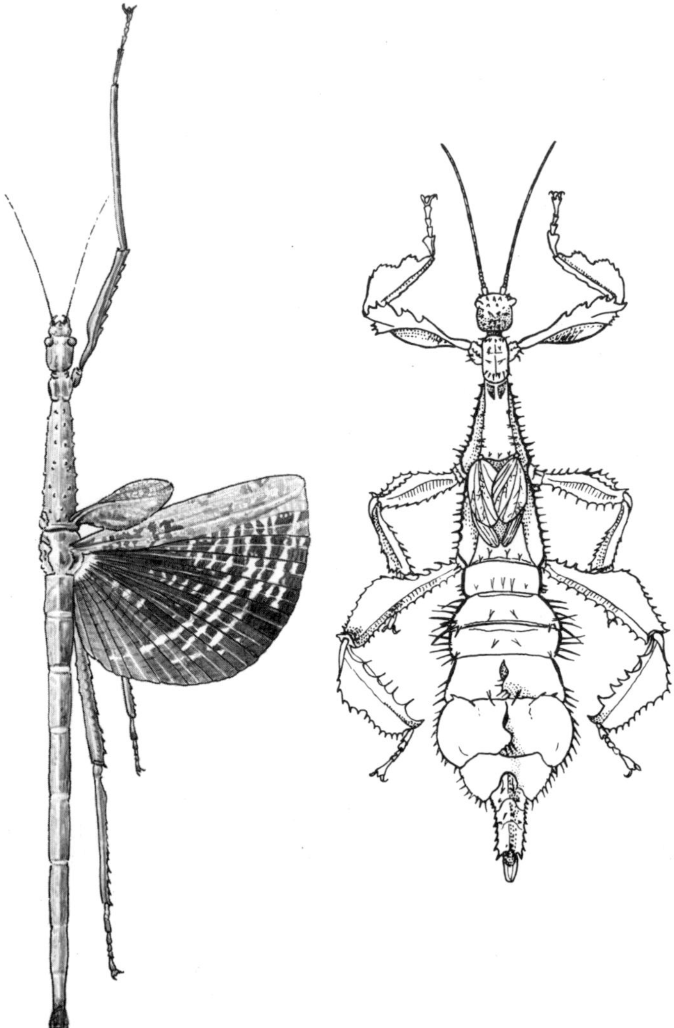

Left: An example of the *Acrophylla* genus, *A. tassellata* female.
Right: *Extatosoma tiaratum*, female.

Stick insects in the family Phasmatidae are generally long and slender. However, those in the species *Extatosoma tiaratum* have a much broader body than other Australian stick insects and the large and numerous expansions and spines on both the body and legs of the female have caused some authorities to name them 'spiny leaf insects'. When resting on foliage, they hang inverted, the abdomen curled toward the back, just like a hanging leaf. The male, however, is less exaggerated in its protective shape and has a much more stick-like abdomen. The female of this species can have a body length of 140mm.

Eggs of stick and leaf insects are laid singly and are usually oval or cylindrical. They have a hard shell and an exterior which may be glossy and deeply grooved. Egg shapes and decorations differ considerably and may be used as a factor in identification of various species.

Females drop their eggs while perched in a tree and in autumn, in certain areas, the constant sound of these eggs falling on litter has been likened to the patter of rain. Once on the ground, the eggs lie all winter, hatching in spring when nymphs emerge. Except that they are smaller, they differ from adults only by their lack of wings, reproductive organs and fewer segments on their antennae.

Once hatched, nymphs make for the upright growing plants and begin eating the foliage. Many die through excessive dryness and others are taken by spiders, birds and ants.

Many stick insects are considered to be rare. The only ones which reach plague proportions are the spur-legged phasmid *(Didymuria violescens)*, the tesselated stick insect *(Ctenomorphodes tessulatus)* and the 'ring-barker' *(Podacanthus wilkinsoni)*.

The defensive behaviour of stick insects rarely includes flight. They rely for protection on their camouflage and their ability to remain rigid. If endangered they will fall from their perch and stay still. If handled they may move wings or legs or release an irritating substance.

Among the 'ring-barkers', differences in colour are found between high- and low-density phases (see Kentromorphism, p. 36). When numbers are low, the ring-barkers are usually green, but at times of high density they are patterned with black and yellow markings. Colour patterns may also be found between these extremes.

WEB- OR FOOT-SPINNERS

Order EMBIOPTERA

Web-spinners are little-known insects with largish heads, noticeable jaws and a short (10mm) dull brown and elongated body. Wings, when present, are found only on males and are smoke-grey with simple venation. Their name and the most conspicuous clue to their presence comes from tubular silk galleries which they weave wherever they go. They are made by silk glands housed in the lower part of the front legs *(tarsi)* and their enlarged tarsi also help to distinguish these insects from termites and earwigs, which they resemble.

The whole anatomy of a web-spinner is adapted to life within its silken tubes. These provide nests and galleries, act as corridors to food supplies and as escape routes from predatory ants. The web-spinning equipment in the tarsus consists of up to 200 silk glands. Ducts channel the liquid from the glands to bristle-like silk ejectors on the under-surface of the leg. The silk is drawn out as the legs brush against a surface and when spinning a new gallery, or repairing an older one, the web-spinner rotates its body while moving the forelegs.

Each tube is spun to suit the individual; those made by hatchlings for instance, which begin to spin before their first moult, are much narrower than those made by an adult. The tube diameter is always narrow enough so that the web-spinner, whether moving backwards or forwards along it, always has its body hairs in contact with all surfaces.

Although web-spinners are not well-known insects, they are found in many places in Australia. They have been found in the high altitudes of the Australian Alps, and on paling

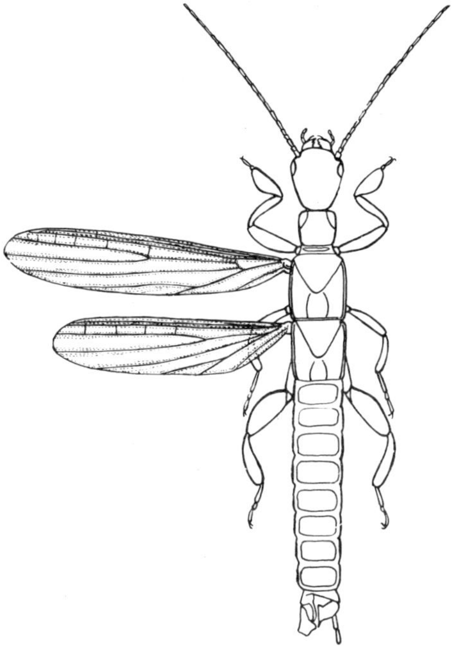

Male web-spinner.

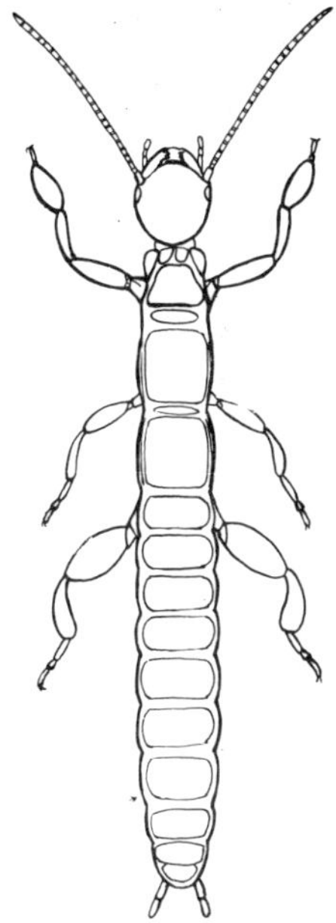

Female web-spinner.

fences in Perth, Western Australia. They build their nests in sheltered locations: under bark, in cracks in dried ground or in old wood.

These insects are wholly vegetarian and live off bark, dead leaves and living moss and lichen. If collected with materials from their native surroundings, they can be reared in captivity in a properly-ventilated glass jar.

At present only 200 species of web-spinners have been described throughout the world, although many more may be present. In Australia, 65 species represent three of eight recognised families. Most identification is based on adult males, with the main characteristics being the head and the structure at the end of the abdomen *(terminalia)*.

BOOKLICE
Order PSOCOPTERA

These are small, soft-bodied insects with an expanded oval abdomen and a length ranging from 1 to 10mm. Their heads are large and possess chisel-shaped jaws from which their earlier name, Copeognatha, was derived. Wings, if present, are held above the body like a tent.

Despite their common name, booklice live mainly on leaves and branches or under fallen leaves or bark, and under rubble. They are often seen on fences and walls, or in caves and can become a nuisance if they multiply excessively in houses, although they do little real damage. Most booklice live on lichens or algae, though laboratory specimens have been reared on yeast culture.

Some species have stridulatory mechanisms on the hindlegs. However, a ticking sound often associated with booklice is caused by the end of the abdomen being struck against the surface on which the insect is standing.

All species seem to possess silk glands and in some species these are used to form webs, in which the community lives. Webs of one genus may be very large.

Before booklice mate, the male performs a mating dance. The female lays her eggs either one at a time or in clusters, on bark or leaves. In some species, males are rare and parthenogenesis (see p. 39) may occur. Spiders, ants, wasps, and other organisms including fungi may predate on or infest booklice and their eggs.

Of the world's 2200 described species, Australia has 120 species of booklice, which are grouped into 21 families and three suborders. Identification of suborders is primarily by the number of segments on the antennae and tarsi. Species of the genus *Liposcelis* contain the traditional 'book louse' which may infest certain homes.

LICE

Order PHTHIRAPTERA

Wingless, with hairy bodies and legs and often with very large, pincer-like claws, lice are uniquely adapted to a life which cannot be maintained without a host. Lice of all species are highly specialised and modified to a particular existence, whether it is eating a host's feathers or hair, or sucking its blood.

Many lice can survive only on a specific form of host and some species are even restricted to a particular part of the host's body. For example, two species of lice are parasites on man. The species *Pediculus humanus* is adapted only to living in the hair of the head while the species *Pthirus pubis* can only exist among the coarser hair of the pubic region. Similarly, three species of lice may infest sheep. One is confined to the face, one to the trunk and one to the legs.

Lice infest most birds and mammals and the only animals which may be free from lice infestation seem to be monotremes and bats, which in turn may be infested by other insects.

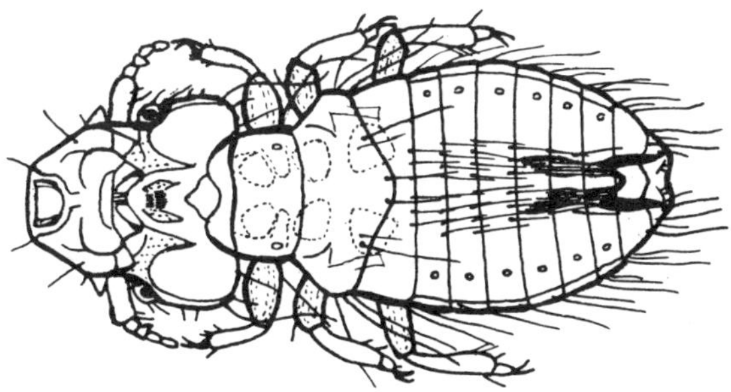

An example of the Philopteridae, the bird-lice family. Length 1.8 mm.

Indeed, the specialisation of lice is such that not only are they averse to light but they can withstand very little variation from the body temperature of the host. If the host dies, lice make their way to the surface of the hairs or feathers, but unless another specific host is nearby, the lice will die either from temperature variations or from the strong light.

Generally lice can only pass from one host to another if those hosts come into direct contact with one another, although some movement may take place between predator and prey, or when lice are transported on other insects, such as flies.

Adult female cattle-biting lice *(Damalinia bovis)* have a life-span of about 40 days, but there is little information on the life-span of other species. Egg laying is frequent and up to 300 eggs are deposited. Death occurs mainly from the host's scratching and preening, though some lice may be cannibalistic.

Lice are classified into two suborders. There are biting lice (Mallophaga) and sucking lice (Anoplura). Biting lice may be found on birds and mammals; sucking lice only on mammals.

Biting lice

Suborder MALLOPHAGA (wool-eaters)

No biting lice are parasites on man, but they are found on many different birds and mammals, and in many cases are highly specialised. Many are hairy and colours include white, yellow, brown, red and black. Their wingless condition is considered to be a fairly recent evolutionary feature and their flattened bodies are of a much more advanced arrangement than those of the true wingless insects (such as silverfish) which are considered not to have had wings in the past.

Eggs of many biting lice are also adapted to suit the host and rarely have the familiar egg-shaped appearance. More often, they resemble stars, flowers and arrows; egg shapes may often be a clue to identification.

Generally, biting lice do not feed on blood but live off the host's feathers or hair. However some species, including the common fowl louse, also take blood by puncturing developing quills or the bird's skin.

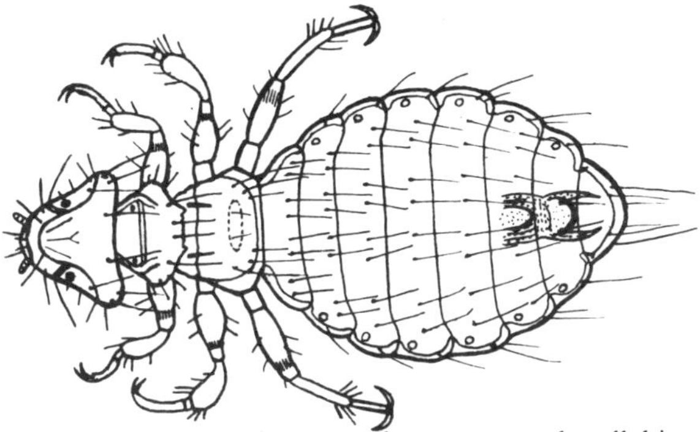

Boopia notafusca, whose hosts are kangaroos and wallabies. Length 1.8 mm.

Although many lice are adapted to a specific host, up to 15 species, representing 12 genera, have been found on a single host species. Some lice have a particularly limited area of parasitisation and Calaby (1970) quotes the genus *Piagetiella* whose species are found only in the inside of pelicans' pouches and on some cormorants.

Most species in the family Boopidae are parasites of marsupials and are known only from Australia and New Guinea.

Most other groups of lice are believed to have arrived in Australia through host species, either with migratory birds or by man and his domestic animals. Introduced species with economic significance include sheep lice, horse lice, cattle lice, goat lice and the widely-distributed fowl lice.

Lice whose hosts are native birds are included in the family Philopteridae and there are at least 90 Australian species including the genus *Dahlemhornia*, which is found on the emu.

Sucking lice

Suborder ANOPLURA (unarmed tail)

Possessing complicated mouths, which consist of sharp piercing stylets which can puncture the host's skin and are then retracted to allow the outer tube to suck the blood, sucking lice are usually smaller than biting lice and rarely exceed a length of 7mm and are often much smaller. Many species have no eyes and in others, the eyes are very much reduced. Their eggs are oval and unadorned and are usually cemented to the hair of the host.

There are five families, which possess different anatomical differences, but they are mainly grouped according to the type of host. They inhabit the hair of only five orders of mammals and for some reason, the cat family, marsupials, bats and other animals are never infested.

The families of sucking lice and their hosts are:

Family	*Hosts*
ECHINOPHTHIRIIDAE	seals only.
LINOGNATHIDAE	ruminants, including cattle, sheep and goats; domestic dogs and dingoes.
HOPLOPLEURIDAE	rodents, including rabbits, hares, rats and mice.
HAEMATOPINIDAE	ungulates, including cattle, buffalo and horses.
PEDICULIDAE	primates, including man and monkeys.

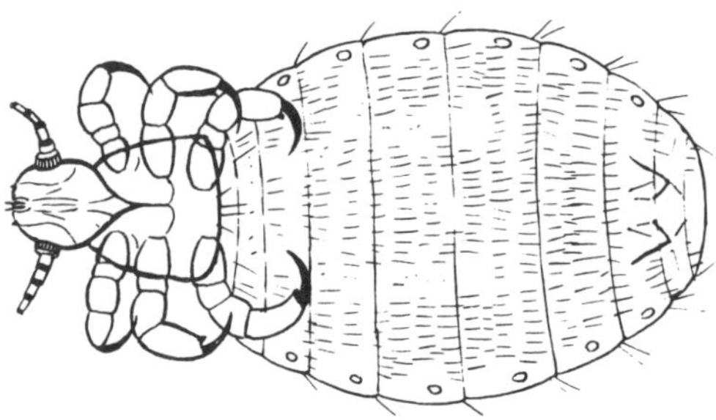

The introduced foot-louse of the sheep, *Linognathus pedalis*. Length 1.9 mm.

To date, some 30 species of blood-sucking lice have been recorded in Australia, and these include species parasitising man and those found on cattle and other domestic animals. It is not known whether the Echinophthiriidae are present, but it is considered probable as Australia has a fair-sized population of seals off its southern coastline.

Hoplopleuridae species are known from introduced rabbits, hares, rats and mice. There are also at least five endemic species which are found on native Australian rodents.

BUGS, LERPS, APHIDS AND CICADAS

Order HEMIPTERA (half-wings)

Possibly no group of insects show such wide variation in appearance as the Hemiptera, with some 4000 species recorded in Australia and is the order with the fourth largest number of species. The appearances of bugs range from the solid, heavy-winged cicada to the slender water-strider, which can skim across the surface of a pond without even ruffling its surface. Some bugs, such as the stink or shield bug, are beautifully coloured while others, such as the plant-hoppers, could superficially be mistaken for moths when in flight with their broad wings extended.

The name 'bug' is an old English word which originally meant any strange creature which was small and, generally, unpleasant. Included in the order are bed-bugs, offensive pests of man. There are others which are significant pests of many native and imported plants.

The term 'Hemiptera' came from certain insects in this order which have only half their front wings thickened to form a hard and horny tegmen. All possess piercing and sucking mouth parts. Metamorphosis of nymphs is gradual.

The order is divided into two suborders—the Homoptera (similar wings) and the Heteroptera (different wings). Their lengths may range from 1 to 90mm.

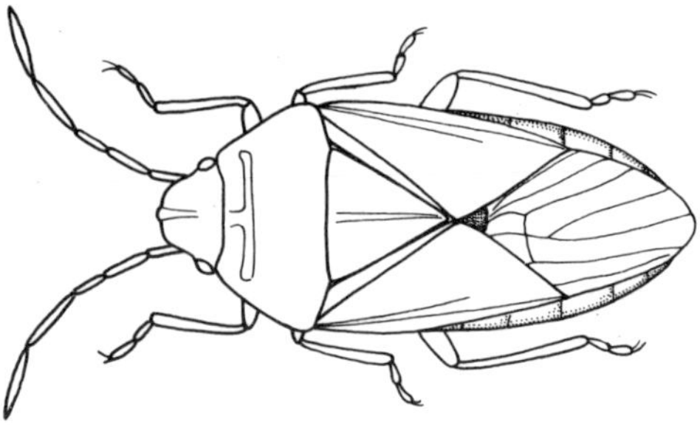

Typical *Heteropteran* bug.

Similar-winged bugs

Suborder HOMOPTERA

In this group, 39 families consisting of 9 distinct types of insects are represented in Australia by more than 2000 species. They have a greater diversity in appearance than any other group of insects. Some species are very small but may nevertheless be capable of causing considerable damage to crops. Others are conspicuous, both for their size and colour.

The main superfamilies are:

Plant-hoppers	FULGOROIDEA
Froghoppers or spittle-bugs	CERCOPOIDEA
Cicadas	CICADOIDEA
Leaf-hoppers	CICADELLOIDEA
Lerps	PSYLLOIDEA
Aphids	APHIDOIDEA
Scale insects, mealy bugs and gall-makers	COCCOIDEA

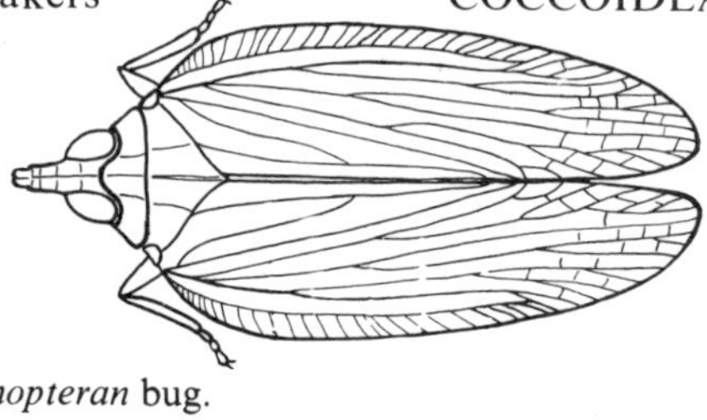

Typical *Homopteran* bug.

PLANT-HOPPERS (Superfamily FULGOROIDEA)

Fourteen families of plant-hoppers are found in Australia; some have broad wings with a span of up to 60mm and are vividly coloured. Most live on flowering plants but some may live on fungi. Several small species native to North Queensland spread virus diseases in cultivated grasses, while others are known to spread diseases in sugar cane, maize and rice.

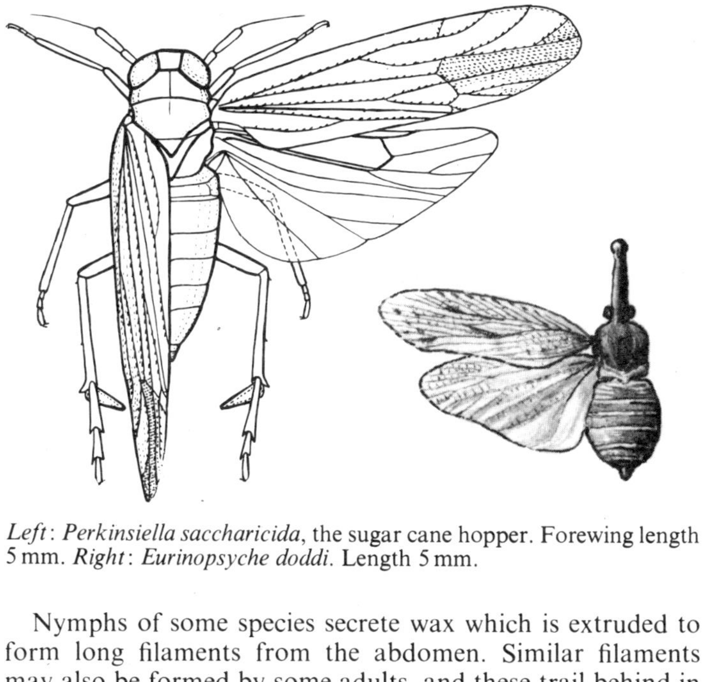

Left: *Perkinsiella saccharicida*, the sugar cane hopper. Forewing length 5 mm. *Right*: *Eurinopsyche doddi*. Length 5 mm.

Nymphs of some species secrete wax which is extruded to form long filaments from the abdomen. Similar filaments may also be formed by some adults, and these trail behind in flight, sometimes 60 to 75mm behind the insect. Their function has not as yet been established.

Some of the family Eurybrachyidae (no common name) have a wing-span of 60mm while lantern-flies (Fulgoridae) have a wing-span of 40mm and may be beautifully marked with crimson and brown. Species of the genus *Eurinopsyche* sometimes have a strangely projecting snout. The name 'lantern-flies' is erroneous and is based on a report by an early Swiss entomologist who claimed that the heads of some insects collected were luminous. However this phenomenon has never been observed again. Some South American lantern-flies also have a strangely deformed and large snout, which has made them the object of superstition among Indians of that area.

The family Achilidae may be coloured in shades between black and green; they fold their thickened forewings flat along the body when at rest. One species, *Achilus flammeus*, is a brilliant red; its nymphs have been found in trees. The families Lophopidae, Ricaniidae and Flatidae carry their wings held over their bodies, like a steep gabled roof. On the species *Lophops saccharicida* (Lophopidae) the front of the head curves forward to give the side view a strangely curved and upstanding appearance.

Passion-vine hoppers *(Scolypopa australis)* in the family Ricaniidae, cause very much damage to passionfruit vines and other plants. The species in the genus *Zoraida* (Derbidae) have very short bodies with wings nearly three times the body length. They are fungus feeders and are found mainly in the tropics.

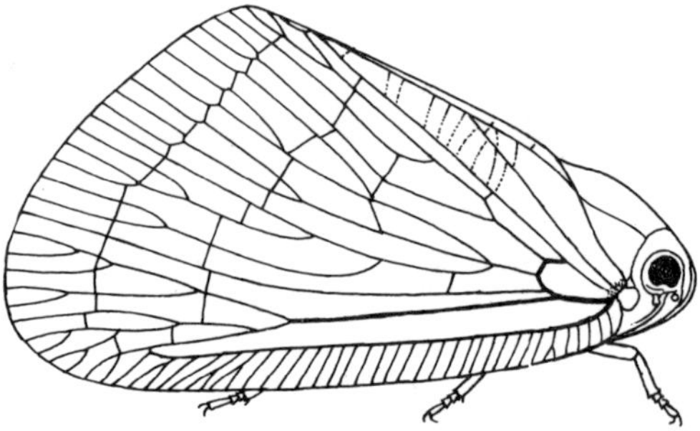

Scolypopa australis, the passion-vine hopper. Length 10 mm.

SPITTLE-BUGS or **FROGHOPPERS** (Superfamily CERCOPOIDEA)

These insects can usually be identified more easily in the nymphal stage than as adults. The youngsters are enclosed in a mass of small bubbles which resemble spittle and are often seen on blades of grass or tender plant stems. If the 'spittle' is carefully removed, the nymph, a pale frog-like insect, can be seen. Usually its beak will be embedded in the plant tissue, sucking the sap. Spittle is produced from the back of the body and by movement of the tail, an abdominal bellows pumps air into the liquid to produce bubbles. As these form, the insect moves them about its body until in a short while, the whole body is covered.

Other nymphs of this suborder live in liquid-filled tubes which are formed as appendages to the host plant. In both cases, the spittle and the liquid act both as camouflage and as insulation from variations in atmospheric temperature.

Adults are not easily identified but some species in northern Australia and New Guinea have brilliant colouring. One of the strangest adult spittle-bugs is the species *Machaerota finitima*, which has the top of its thorax *(scutellum)* formed into a backward-pointed crest.

All froghoppers possess two compound eyes and, in addition, have two simple eyes *(ocelli)* above the head.

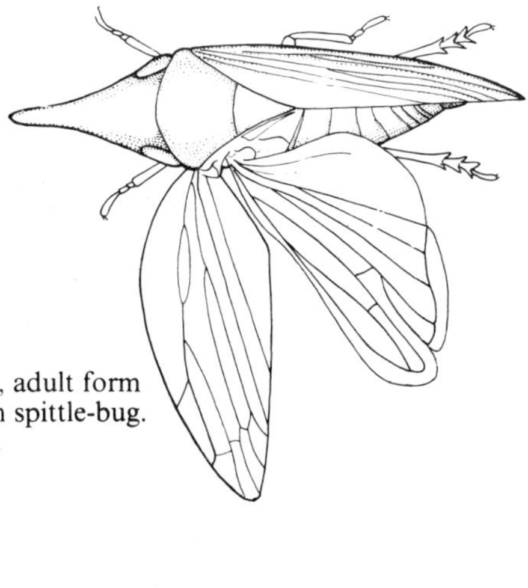

Philagra parva, adult form of the common spittle-bug. Length 10 mm.

CICADAS (Superfamily CICADOIDEA)

Sometimes erroneously named 'locusts', cicadas are best known for the communal chorus of 'drumming' which they produce with special stridulatory mechanisms (see p. 29). The female lays her eggs in incisions she has made in branches and twigs with her sharp ovipositor.

When the eggs hatch into nymphs a few weeks later, they drop to the ground where they burrow into the soil and attach themselves to a tree rootlet. The nymphs spend considerable periods in the ground, feeding on the sap of the roots until fully grown. In one American species, this subterranean part of its life-cycle is said to last up to 17 years. Australian cicadas may spend several years in this stage of development. When the nymph is fully grown, it climbs the nearest tree, fastens its claws into the bark and sheds its outer skin.

Australian cicadas are grouped into two families. The Tettigarctidae are found only in south-eastern Australia, where the two living species are descendants of cicadas known from fossils found in Mesozoic and Tertiary rocks (from 60 to 200 million years old). They are conspicuous for their hairy bodies.

The other 195 species of cicada in Australia are included in the family Cicadidae, in which the females lack the tymbals which are part of the stridulatory mechanism.

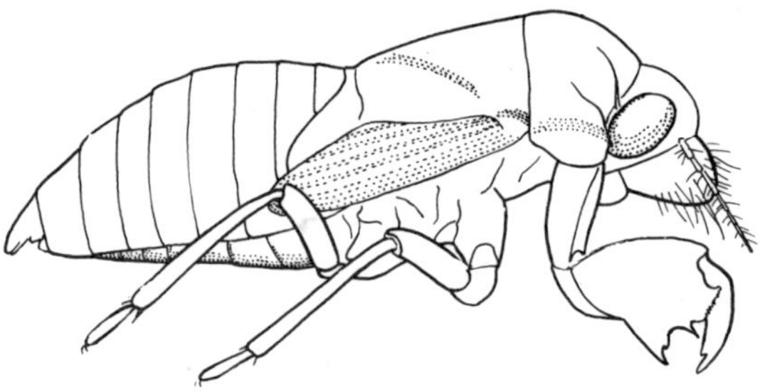

Burrowing cicada larva. Length 27 mm.

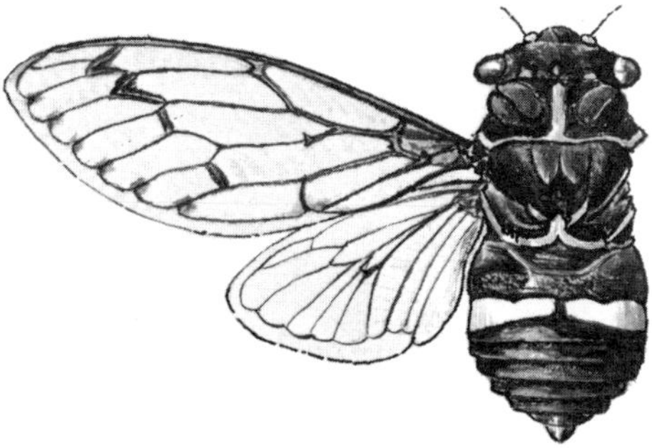

Henicopsaltria fullo, the white-banded cicada of Western Australia. Body length 30 mm.

Among the best-known Australian species are the 'double drummer' *(Thopha saccata)* with a 12.5cm wing-span and the green Monday *(Cyclochila australasiae)*. A dusky pink variety may also be found in some areas.

Cicadas have normal compound eyes and three simple eyes on top of the head.

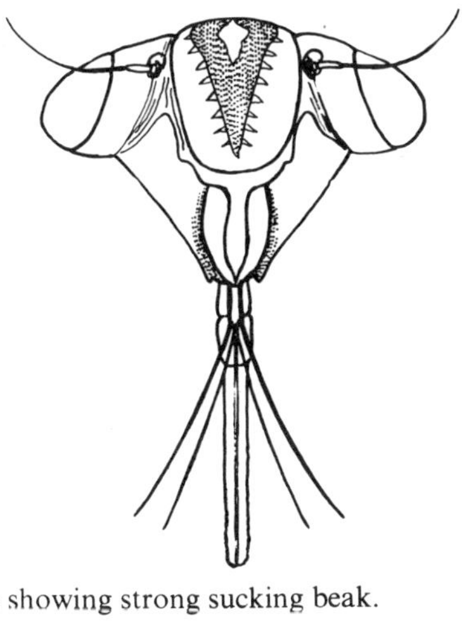

Head of cicada showing strong sucking beak.

LEAF-HOPPERS (Superfamily CICADELLOIDEA)

In Australia leaf-hoppers are represented by three of the five families in the world and of these, 500 of the 650 described species are included in the family Cicadellidae. Best described as 'torpedo-shaped', these insects are found in a variety of body dimensions. While some have elongated and rounded heads, others have a much squarer head shape when viewed from above, with an overall wedge-shaped appearance. The introduced apple leaf-hopper *(Typhlocyba froggatti)* is an established pest in apple orchards. The most common and possibly best-known leaf-hoppers are found in the genera *Eurymela* (gumtree hoppers) and *Eurymeloides*, these insects having vivid white, yellow and red markings on their deep blue or black bodies. Leaf-hoppers are often found in large numbers and when disturbed they fly away rapidly, or move behind the leaf stems and hide from the intruder. Sometimes they hop from one stem to another and it is this behaviour which has earned them their common name.

Many species tend to feed on the young sappy growths of *Eucalyptus* trees and they are often attended by ants who are attracted by the 'honeydew' excreted by leaf-hoppers.

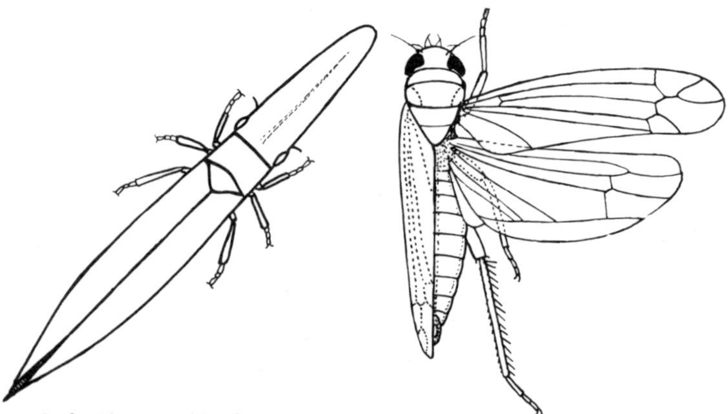

Left: Elongated leaf-hopper. Length 11 mm.

Right: The apple leaf-hopper. Length 3 mm.

TREE-HOPPERS (Family MEMBRACIDAE)

These insects are conspicuous for their swollen and often grotesquely-shaped heads (really prothorax or pronotum lying between head and wings) which may possess swollen or expanded 'cheeks'. The thorax may often be developed into 'horns' or other strange shapes. The best known species is the green coloured black-wattle feeding *Sextius virescens*. Others may be black or brown. Average length is about 10mm. Like the leaf-hoppers, tree-hoppers are communal, often being found in large numbers on particular trees. Tree-hoppers lay their eggs in slits cut in the plant tissue and the insertion of many eggs can often cause as much damage as the sap-sucking by adult insects.

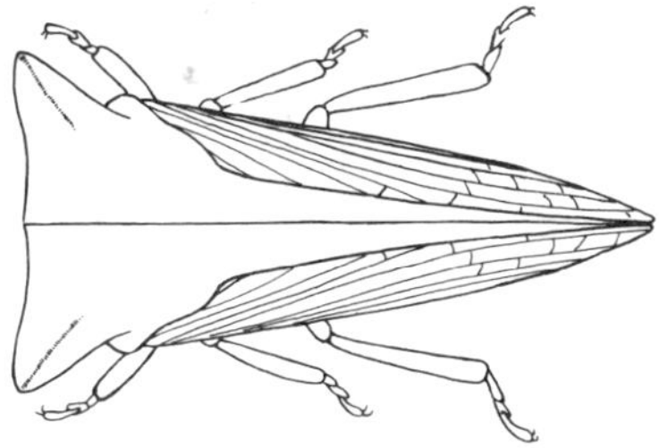

Sextius virescens, the green tree-hopper. Length 9 mm.

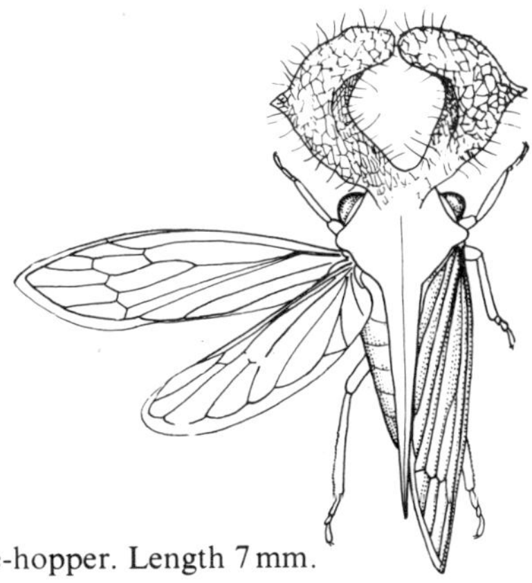

Lubra regalis, a horned tree-hopper. Length 7 mm.

LERPS or **LERP INSECTS** (Family PSYLLIDAE)

Represented in Australia by some 300 species most lerps seem to be associated with particular host plants or a closely related group of plants.

The name 'lerp' comes from the characteristic scale or lerp which the larvae of these tiny insects spin on leaves or small twigs. The larvae shelter under the lerp, which is produced from an anal discharge and spun into a protective shield, often of distinctive appearance. One of the most conspicuous of these structures is of the lace lerps in the genus *Cardiaspina* which are complex. Often they are flat and lacy but those of *C. textrix* are like a cage of fine golden threads through which the larva is clearly visible. The well-known sugar lerp (*Spondyliaspis eucalypti*) weaves a white conical lerp. These are sometimes thickly massed on eucalypt leaves and when they fall to the ground, aboriginal and bush children often suck them for their sweetness, which is nicknamed 'manna'. Aborigines also steep them in water to obtain a sweet drink.

The lerp of *Lasiopsylla rotundipennis* looks like an irregular drop of candle grease. The shape of *Rhinocola costata* is like a scallop shell; that of the ironbark or spotted gum psyllid (*Eucalyptolyma maideni*) is like a miniature fern leaf and *Spondyliaspis rubra* has a bright scarlet lerp. Some psyllids (genus *Glycaspis*) and the genus *Lasiopsylla* form spherical galls which have a hole at the base which is plugged with wax.

Lerps are usually associated with the curling of eucalyptus leaves, which may provide additional protection against the larva's dehydration. The damage caused by lerps has yet to be fully assessed. Some in plague numbers may be responsible for total defoliation of a tree but more often they may damage parts of the tree without actually killing it.

The fig psyllid (*Mycopsylla fici*) lives on the leaves of the Moreton Bay fig in Queensland. These insects cause the rubbery sap to exude and when the leaves fall, they stick to the feet of passing animals and humans.

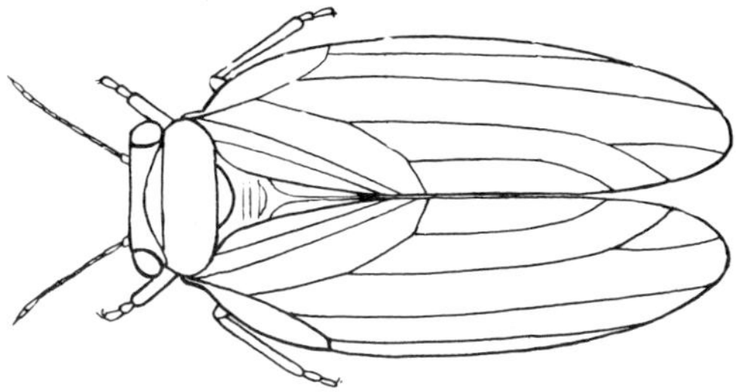

Lerp insect. Length 2 mm.

APHIDS, GREENFLY, BLACKFLY or **PLANT LICE**
(Family APHIDIDAE)

These are notorious enemies of home gardeners and although Australia has only 100 described species of the world's total of 3600, they are capable of causing much damage. Aphids are responsible for two forms of damage. They directly cause leaves to wilt by sap-sucking; some species also carry plant virus diseases which they transmit.

Like lerp insects, many aphids have a wax-secreting gland and one of the best known natural history examples of mutually beneficial partnerships *(symbiosis)* between insects is that of ants and aphids—which are sometimes popularly known as 'ant cows'. The aphid's secretion, like that formed by leaf-hoppers and lerps, is sweet and is also known as 'honeydew' because it falls on leaves in dew-like droplets. This is gathered by ants, and, in some seasons, by bees too.

In the course of evolution, particular species of aphids and particular species of ants have become mutually reliant *(commensal)*, and in one example, the ants not only gather the 'honeydew', but also gather the aphid eggs, moving them to their underground chambers during winter. When the eggs are about to hatch the ants carry them to the roots which the aphid nymphs will suck. When they are feeding in exposed locations, these ants will also protect their 'herd' of aphids against predators such as aphid lions and ladybirds. The ants may also build shelters over the aphids and move them to new locations of food.

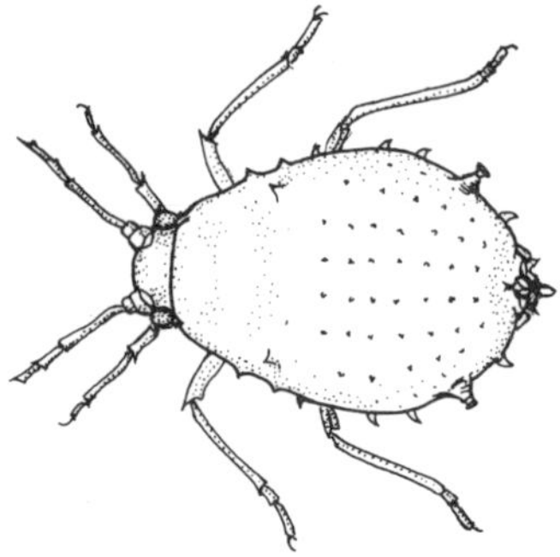

Aphid.

Experiments in rearing aphids has shown that a single species, if fed on different plants, can develop an entirely different appearance, just as those reared at high temperatures may be paler and have less hairs than the same species reared in cooler conditions. Females of different groups may produce live young in a winged or wingless state, or they may produce eggs. Greenfly and blackfly are aphids well-known for living in exposed locations. Many others exist in unfolded or crumpled leaves or in inconspicuous, near-to-the-ground locations. Most of Australia's aphids seem to have been introduced accidentally, either by the agency of man or possibly by wind transport.

A wide variety of plants is attacked by aphids, including palms and orchids, which are hosts for three species in the genus *Cerataphis*.

One of the most damaging aphids is the root-sucking phylloxera *(Viteus vitifoliae)* which, in 15 years between 1895 and 1910, wiped out nearly all Australian vineyards and put an end to the first wine industry. The ravages of this insect were only rectified by grafting European vines on hardier phylloxera-resistant root-stock from America. Probably the best known species of aphids are the bright green or pinkish rose aphis *(Macrosiphum rosae)* which swarm on young buds and shoots of roses. Another species transmits 'bunchytop disease' in bananas in tropical regions. More recent problems have been caused to lucerne by the alfalfa and blue-green aphids, both from the U.S.A.

SCALE INSECTS, MEALY BUGS and **GALL-MAKERS**
(Superfamily COCCOIDEA)

Like the groups already described, the superfamily Coccoidea contains minute insects which may be grouped into three distinct types. They are: scale insects which cover themselves with a waxy 'scale'; mealy bugs which are covered with waxy, flour-like down; gall-makers which cause swellings in plant leaves and stems.

Australia has some 520 species of these insects and identification, in comparison with aphids and lerps, can only be carried out under magnification. It is based on wing venation, presence of leg tarsi, and the number of segments on the antennae.

Males are minute-winged insects; the larger and fatter females are wingless and, in many species, lose both antennae and legs; larvae are active six-legged creatures with little resemblance to the adults of either sex. Most of the life-cycle is spent sucking sap from beneath a protective covering.

One of the most damaging scale insects is the citrus red scale *(Aonidiella aurantii)* which causes widespread damage to citrus crops. These larvae are covered by a waxy secretion, left by the female to protect her eggs. The Mexican cochineal insect *(Dactylopius coccus)* is a scale insect which feeds on cactus. The females of this species are dried and 70 000 insects are used to make 500g weight which is processed to produce a rich red dye, now used to colour food, cosmetics and drinks. The species was introduced into Australia in an attempt to control prickly pear, however this was finally achieved with *Cactoblastis cactorum* (Lepidoptera). The family Lacciferidae is represented in Australia but not the Indian lac insect *(Laccifer lacca)*, whose wax is processed to make shellac.

Mealy bugs are large soft, flat insects with naked bodies and although covered with the dusty covering of wax, do not form a scale. Some are pest species, affecting pineapples and citrus trees.

Some of the largest mealy bugs are in the genus *Callipappus*, with females up to 25mm long and blue-black in colour. The males have a completely different appearance, being smaller with the tip of the abdomen having a tuft of fine silvery thread which has earned them the common name 'bird of paradise flies'.

Gall-makers live within the plant tissues. The 'galls' are formed by the larva attaching itself to the bark of a twig or leaf. Their presence causes an irritation which causes the plant tissues to expand and cover the larva until it is completely enclosed. Females remain imprisoned in the gall but males make an opening in the gall and escape.

Some gall insects make distinctive galls. Those in the genus *Cylindrococcus* make capsule-shaped galls on casuarina trees, and that of *C. spiniferus*, which also inhabits the casuarina, makes a gall very similar to its host tree's seed vessel.

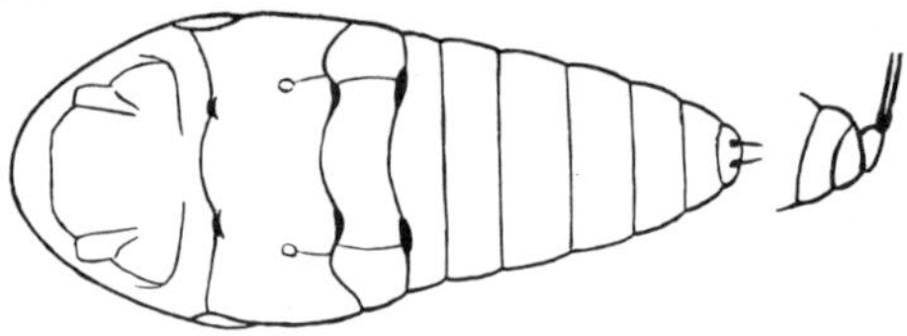

Apiomorpha duplex, female, a gall-maker on eucalypts.

Gall of *Cylindrococcus spiniferus*.

Different-winged bugs

Suborder HETEROPTERA

There are 48 families in this group, representing 1577 described species in Australia. There are nine distinct groups with common names, and five others. The eight principal aquatic groups are considered to be superfamilies. The latter are discussed and described on page 214 under 'Aquatic Insects'.

The insects in this group usually have forewings which develop into hemelytra, or a hard wing-case similar to that of a beetle.

The suborder is usually divided into three sections:

HYDROCORISAE—with short antennae mostly living in or associated with water.

AMPHIBICORISAE—which are mostly adapted to life on the surface of water.

GEOCORISAE—which are mostly terrestrial.

The main terrestrial superfamilies are:

Bed-bugs and bloodsuckers	CIMICOIDEA
Lace bugs	TINGOIDEA
Assassin bugs	REDUVIOIDEA
Shore bugs	SALDOIDEA
Flat and bark bugs	ARADOIDEA
Seed-eating bugs	LYGAEOIDEA
Shield or stink bugs	PENTATOMOIDEA
Negro or burrowing bugs	CYDNIDAE (Family)

BED-BUGS and **BLOODSUCKERS** (Superfamily CIMICOIDEA)

These are undoubtedly the form of Hemiptera most directly concerned with man. In the family Cimicidae, the only species recorded in Australia is the international bed-bug *(Cimex lectularius)* which has been introduced since European settlement. Outside Australia, there are other species in this family which also suck the blood of birds and bats, but to date these have not been recorded in Australia.

Like lice, bed-bugs are associated with the presence of man and shelter in cracks in walls and furniture during the day. At night they emerge to feed on the blood of humans and some animals. Fortunately, the advent of DDT and other methods of fumigation and cleaning have considerably reduced the incidence of these pests, as will eventual demolition of unhygienic dwellings in some of the poorer parts of the older cities and towns.

In appearance, the bed-bug is flat, bristle covered, red and squat. It is wingless and its mouth is of the piercing-sucking type.

Other Australian bloodsuckers in the superfamily Cimicoidea include small parasites of bats and the family Anthocoridae which live on insects or their eggs.

One of the largest families in this group is the Miridae (sometimes known as leaf or plant bugs) which contains both blood-suckers and sap-suckers. These bugs are winged, with small heads and long bodies. Most are green or brown, and some may be patterned. Some species attack crops. Among the 100 described species in this family, there are five species in the Orthotylinae subfamily which are ant mimics. This form of imitation is known as 'Batesian mimicry', a term which describes a group of insects which is neither distasteful nor injurious to a predator but its appearance, form and behaviour are similar to a different insect which is better protected.

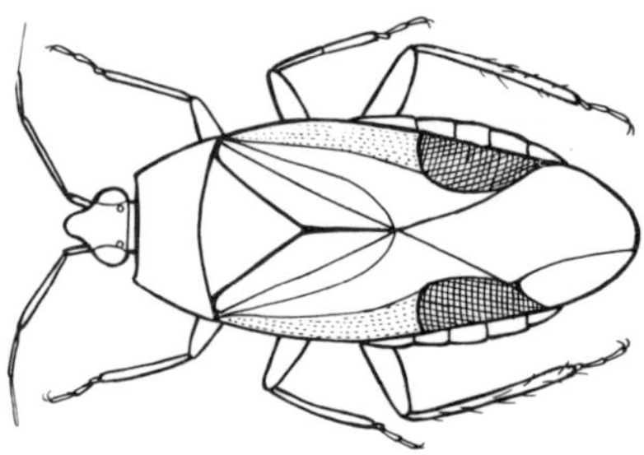

A member of the Anthocoridae, *Lyctocoris campestris*. Length 4 mm.

Two species of the Phylinae subfamily in south-western Australia are able to exist on the sticky leaves of the insect-eating sundew plants and live by sucking the blood of other insects trapped in these plants.

LACE BUGS (Superfamily TINGOIDEA)

So named because of the lacy appearance of the raised veins on their hardened forewings, some Australian species are said to damage commercial and garden plants, including azaleas and olive trees. The central American lantana bug *(Teleonemia scrupulosa)* was introduced to Queensland to attack the lantana plant. Lace bugs rarely exceed a length of 5mm, are poor fliers and have prominent transparent raised projections above the head and thorax.

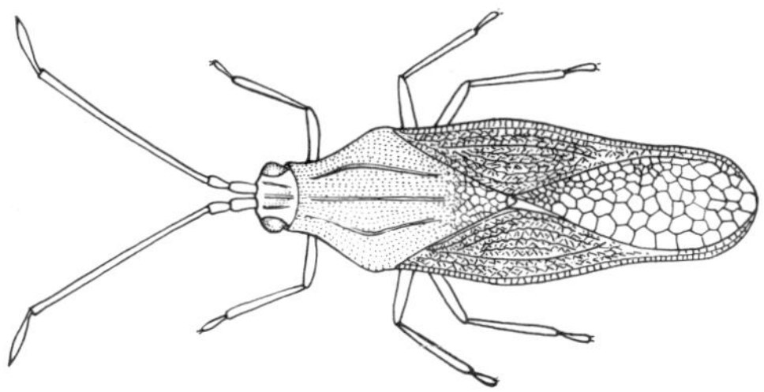

The lace bug *Froggattia olivina*, a pest of olive trees. Length 3 mm.

ASSASSIN BUGS (Superfamily REDUVIOIDEA)

Largely predators of other insects, there are about 240 described species of assassin bugs in Australia. None of these are known to attack man but several of the larger species can inflict a painful 'bite' or sting with the poison they inject. Among the best known assassin bugs are the bee killers, which may be 25mm long, in the species *Pristhesancus papuensis*. They are predominantly brown in colour. Others are reddish-brown and black, some with spines on the head and pronotum.

While some assassin bugs are long and mantid-like, others are flattened. One feather-legged bug (genus *Ptilocnemus*) deposits a secretion which attracts and anaesthetises the ants on which it preys. Many assassin bugs have brilliant red, black or yellow and black markings. Other species emit an offensive odour to deter predators.

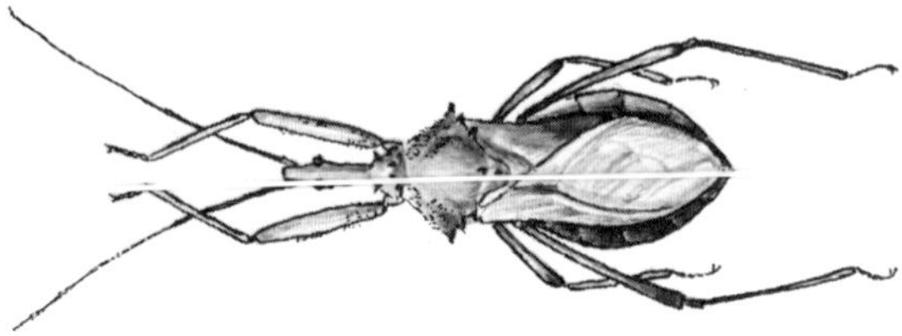

Pristhesancus papuensis.

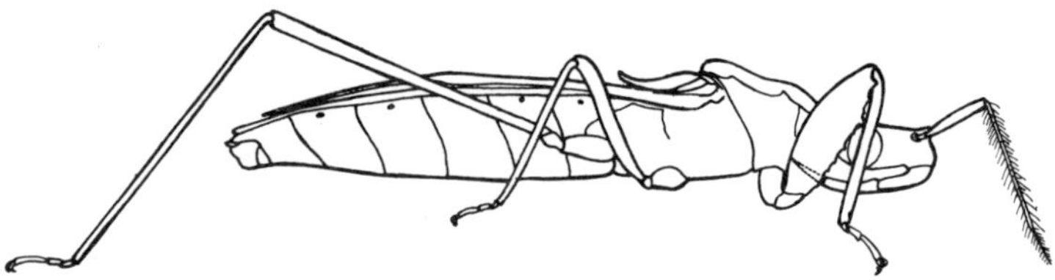

Mantid-like assassin bug, *Oncocephalus confusus*. Length 17 mm.

SHORE BUGS (Superfamily SALDOIDEA)

These are dull-coloured insects with an oval head and body; able to run rapidly, they can also jump, and move in a series of short but sudden flights near the ground. They live mostly near the water's edge in freshwater and seashore locations. A small beetle-shaped species, *Omania marksae*, is found on the Great Barrier Reef and at high tide retreats into cracks in the coral.

FLAT, BARK and **FUNGUS BUGS** (Superfamily ARADOIDEA)

As their names indicate, these are flattened and adapted to a life beneath the bark of trees. Their antennae have six segments. They are often found beneath the bark of an old log, although occasionally they may be seen slowly moving along outside the bark. Their irregular shape and dark brown or black colour provide excellent camouflage against rough, moss-covered bark. It is believed that the female remains close to her eggs and sprays them with a repellent scent to drive away predators. These bugs are believed to feed almost exclusively on fungus. The young are red and somewhat similar to bed-bugs. Their presence in freshly sawn timber has caused the erroneous impression that bed-bugs are found outside houses and away from human settlements. However, these tiny red bark-bug larvae are not bloodsuckers. An interesting family within this group is the Termitaphididae, which are wingless and have been found inside termite mounds.

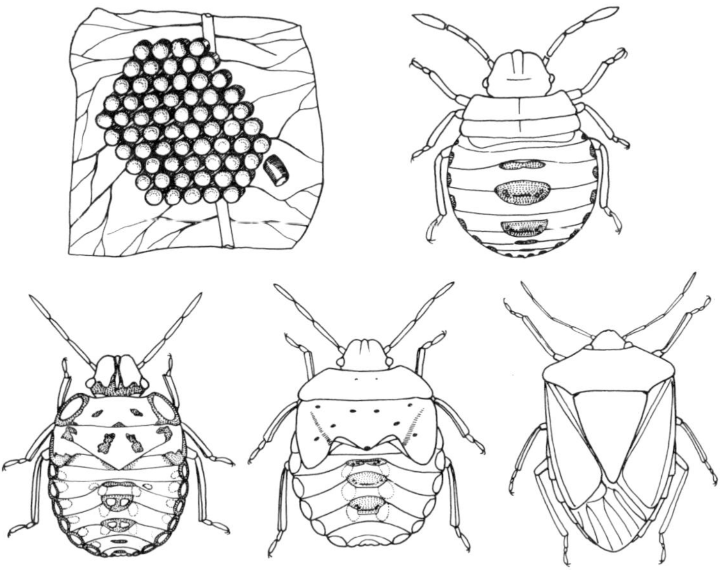

Life history of green vegetable bug *Nezara viridula* (see shield bugs p. 69) from egg to adult through three larval instar stages (skin castings), showing development of shield, wing buds and abdominal scent glands.

SEED-EATING and **CHINCH BUGS** (Superfamily LYGAEOIDEA)

Represented by 230 described species in Australia, these bugs vary in size and habitat. One of the best known is the Rutherglen bug *(Nysius vinitor)*, which occurs in many parts of the country and is a common pest of fruit and vegetables. Although small, these minute grey insects, only 3mm long, occur in large numbers. Their eggs have been found on sow thistles, cudweed and blady grass and in laboratory tests, one female laid 435 eggs.

Similar in size though conspicuous with black and white markings, the cotton-seed bug *(Oxycarenus luctuosus)* is known from inland Australia and has been recorded as a pest on cotton and hibiscus plants. In dry seasons, this species or related ones may also damage orchards. The chinch bug of America has done tremendous damage to corn and other grains and is represented in Australia by eight species which are poorly known and evidently cause little damage.

In the family Lygaeidae, the genus *Clerada* is reported as living as a parasite in the fur of small mammals. The strange long-legged stilt bugs (family Berytidae) are plant-suckers.

Damage to cotton crops is caused by the red-cotton or pale cotton-staining bug *Dysdercus sidae*, which stains the lint. It has a yellow-brown body and a red head. Related species affect deciduous fruits. These colourful bugs are also known as harlequin bugs (family Pyrrhocoridae) or firebugs, as a result of their red colouring.

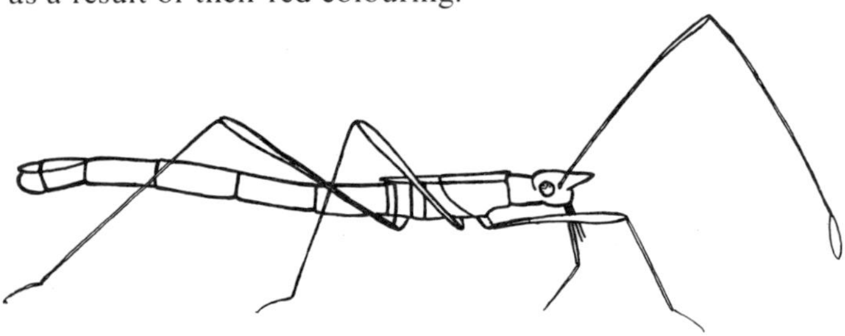

Stilt bug. Length 6 mm.

SHIELD and **STINK BUGS** (Superfamily PENTATOMOIDEA)

With 459 species recorded in Australia, these plant-sucking bugs are still classified in different ways by different authorities. There is however universal agreement that the common names given are well suited. The shield bugs all have a large shield-shaped part of the mesothorax *(scutellum)* and often this is brightly coloured. Well-known species include the bronze orange bug *(Musgraveia sulciventris)* a pest of citrus trees, and the green vegetable bug *(Nezara viridula)* an introduced pest which infests tomatoes, beans and lucerne.

Shield bugs in the subfamily Asopinae have a strong beak which is used to impale caterpillars whose flesh they suck. If disturbed the bug seldom abandons its prey, but moves away carrying the caterpillar in front of it.

Also included in this group are the burrowing or negro bugs in the family Cydnidae. These are usually a shiny black or brown and live in the soil, but they may be attracted to light. Their hind-legs have spines for digging and they burrow rapidly into the soil when disturbed. These bugs may be confused with some beetles and the scutellum or back plate must be examined to see if it is in a single piece. In black beetles of similar appearance, there is a split down the centre forming two *elytra* or wing-cases.

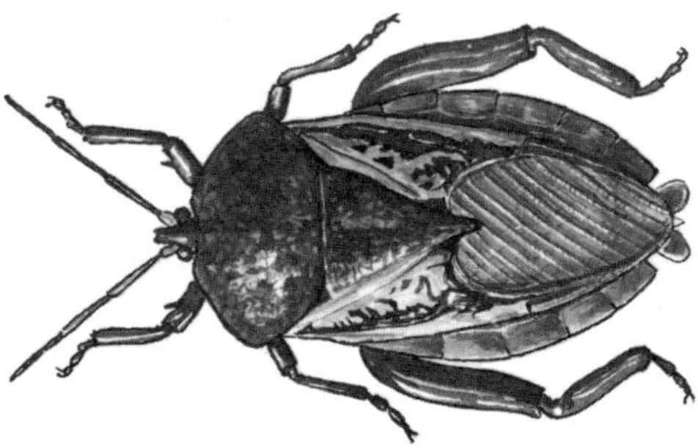

The largest of the Australian pentatomids, *Oncomeris flavicornis*. Length 45 mm.

THRIPS
Order THYSANOPTERA (fringe-winged)

Thrips are small insects, usually less than 12mm long, best known for the damage some species cause to garden plants and crops by sucking the sap. They have elongated, soft bodies and two pairs of slender wings, fringed with long hairs; it is from these wings that the order gets its name. Their feet are clawless and often end in suckers which allow them to stick to smooth surfaces. Their mouth parts are adapted to scraping and sucking. Some species of thrip are parthenogenetic (see p. 39). Nymphs pass through two or three resting stages between moults.

Although some 4000 species of thrips are known throughout the world only 400 species have been described in Australia, and many more may in fact occur. Thrips are grouped into two suborders: Terebrantia contains species which tend to favour the sap of plants and flowers, while those in Tubulifera tend to live in debris and may feed on fungi. Some species may be predatory, living on small insects and mites.

Thrips are not strong fliers but may be carried considerable distances by strong winds. Their numbers seem unaffected by such predators as wasps, bugs and spiders and only the weather seems to determine if they occur in pestilential numbers.

Thrips may cause surface blemishes which result in flowers turning brown. They may also damage the sexual organs of fruits and vegetables and prevent formation of the crop. Thrips attack young buds and ripe fruit and may also spread plant viruses and fungi such as the tomato spotted wilt and wheat rust.

Australia's thrip population includes many species present elsewhere in the world and also some that are unique to this area. One of the world's largest thrips is the long-headed Australian giant thrip (*Idolothrips spectrum*) which is found in the warmer regions, usually on decaying eucalypt limbs.

Gall-forming thrips in the genus *Kladothrips* are found only in Australia and Indonesia. They form galls on acacia trees (wattles), often in the interior. It has been suggested that gall-forming may be a method of survival against the dehydrating effects of sun and winds in these areas. Another species in this genus develops huge bubbles in the first three abdominal segments. These contain egg masses.

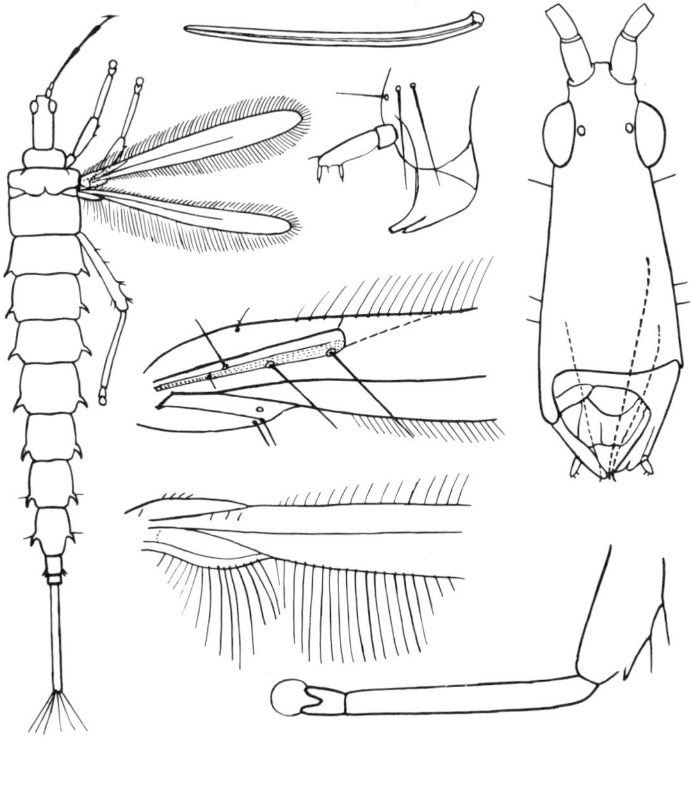

Idolothrips spectrum, the giant thrip. Length 12 mm.

ENDOPTERYGOTE ORDERS

These insects develop wings internally during larval life and there is a true pupa or resting stage.

Alderflies and Dobson flies	MEGALOPTERA
Lacewings and antlions	NEUROPTERA
Beetles	COLEOPTERA
Stylops	STREPSIPTERA
Scorpion flies	MECOPTERA
Fleas	SIPHONAPTERA
True flies	DIPTERA
Caddis flies	TRICHOPTERA
Moths and butterflies	LEPIDOPTERA
Sawflies, wasps, bees and ants	HYMENOPTERA

ALDERFLIES AND DOBSON FLIES

Order MEGALOPTERA (large-winged)

Found only in the cooler climates and in mountainous regions, Australia has a limited Megaloptera population. Only two species of alderflies and 14 Dobson flies have been described. These are found only in the upper reaches of east-coast streams, in the mountains of the south-east, and in Tasmania. One Dobson fly is known from the south-west of Western Australia.

Adult Megaloptera have a moderately long, dark-coloured body with longish antennae and a soft abdomen. Their semi-transparent wings are broad and wide, with conspicuous dark veins. At rest, the wings are folded obliquely against the sides of the body. Their flight is sluggish and activity is confined to times when the light is less intense.

Megaloptera rest on trees and shrubs near the water's edge and flights are often short. When over water, they often glide. These insects alight on plants near water to mate. Subsequently, females lay from 1000 to 3000 eggs in clusters, choosing shady locations on prominences overhanging the water. Many Megaloptera females may come together to lay and often large surfaces are found covered with the eggs. On rocks, these egg masses seem like lichens and egg colours vary from pale white or brown when deposited to black as they darken during incubation.

Nymphs hatch in two to four weeks and drop to the water below. (See 'Aquatic Insects' pp. 227–8.) Birds and frogs predate on these insects; beetles also eat the egg masses. When adults dip onto the water during flight, they may be eaten by fish and in some places, form a substantial part of the fish's diet.

Dobson flies are grouped in the family Corydalidae and have a wing-span ranging from 70 to 170mm. They have three simple eyes *(ocelli)* and a simple fourth segment on the tarsi.

Alderflies in the family Sialidae rarely have a wing-span exceeding 24mm. They have no simple eyes and have a split fourth segment of the tarsus.

LACEWINGS AND ANTLIONS

Order NEUROPTERA (nerve or net-winged)

The adult and nymphal stages of some lacewings are so distinctive that many people think they are different insects. Adults are generally large with gauze-like wings. The larvae of the family Myrmeleontidae are the well-known antlions, which most people do not associate with such harmless adults.

Possibly the group which best typify the adults are the 'golden-eyes' or green lacewings (family Chrysopidae). They have large transparent wings which are held like a roof above the body when at rest. The wing veins are numerous and are often divided by numerous cross veins, while the main veins end in twigging. This family has wings of a delicate green shade and a slender green body. The antennae are long and on some species, the eyes have a golden shine.

Green lacewings are often found in gardens and homes and some, when disturbed, will release a pungent fluid. This has earned them the alternative name of 'stink-flies'. Adults are

Green lacewing, *Dictyochrysa fulva*. Wingspan 40 mm.

often abundant on flowering eucalypts.

E. F. Riek (1970) writes that with the exception of the Hymenoptera (see pp. 193–213) there is no order of insects whose members are so beneficial to mankind as the Neuroptera. Both as larvae and as adults, lacewings feed on many sap-sucking insects and also on mites and butterfly eggs. They also feed on the egg capsules of spiders.

Parasitic wasps however infest larvae of lacewings and cocoons may be infested by beetle larvae. Bats may eat the adult insects, as will large ants, spiders and frogs.

Another distinctive feature of this family is the stalked eggs of some species which are laid in rows, or at random, on a leaf. When laying these distinctive eggs, the female first exudes a secretion which hardens on contact with the air. This forms a fine thread or stalk, two or three times as long as the white or green egg which is placed at its tip. The eggs hatch in seven to 14 days and the carnivorous larvae live in vegetable debris on the ground. When the time comes to pupate, the larvae spin a silken cocoon, which is usually made in a crevice and covered with insect or vegetable debris.

The Neuroptera are represented in Australia by 396 species in five superfamilies and 15 families. Considerable variations are found between adults in different families and these may be summarised as follows (with the exception of Chrysopidae, already described).

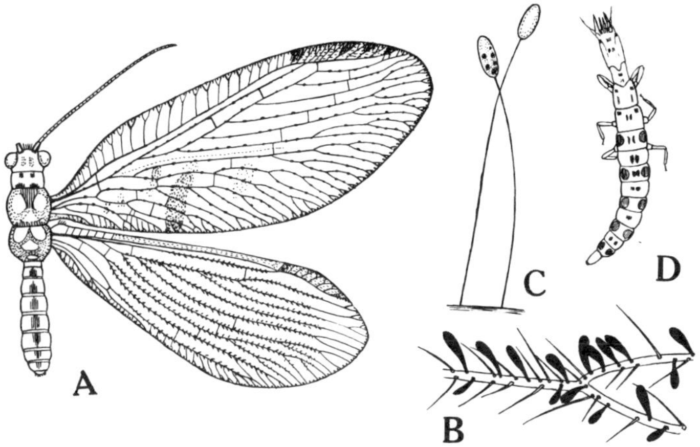

(A) Lacewing of the Berothidae, showing also (B) hard, seed like scales distinctive of Berothidae on hindwing enlarged 60 times, (C) stalked eggs of the family, (D) larva.

MOTH LACEWINGS (Family ITHONIDAE)

Large insects with thick bodies and a wing-span up to 65mm. Their eggs are laid in sandy soil. They are found in low-lying sandy areas near the coast and in the higher country of eastern New South Wales and Queensland.

MANTID LACEWINGS (Family MANTISPIDAE)

Growing to a wing-span of 50mm, these lacewings are remarkably like mantids. Their forelegs are adapted for grasping, the neck is extended and the head is triangular. Some species have brightly patterned wings. On a species of Calomantispa, the base of the wings is bright red and black. Their tiny eggs on short stalks are deposited on trees in groups of several hundred. The larvae live beneath bark.

BROWN LACEWINGS (Family HEMEROBIIDAE)

These are small, fragile insects with very long tapering

antennae. They are found mainly in coastal and mountainous areas of eastern Australia. If disturbed in daylight, Hemerobiidae pretend to be dead.

SILKY LACEWINGS (Family PSYCHOPSIDAE)

Large insects with a wing-span of up to 60mm and short antennae. The fine hairy wings are often beautifully marked. These insects are mainly nocturnal and eggs are laid in *Eucalyptus* trees. They are found mainly in Queensland and New South Wales.

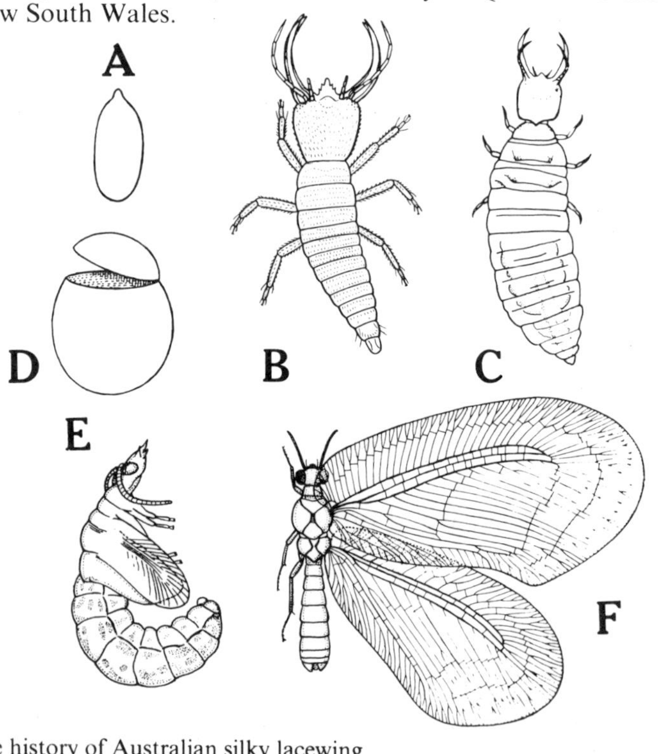

Life history of Australian silky lacewing.
A—egg; length 1 mm. B—first larval stage, length 2.7 mm. C—last larval stage, length 12 mm. D—cocoon. E—pupa removed from cocoon. F—imago, 18 mm.

Family **STILBOPTERYGIDAE** (no common name)

Found only in Australia and South America, these include some of the largest of the lacewings with a wing-span reaching 115mm. Both wings and body are narrow and they are found in both the eastern and western parts of Australia. Larvae may be like antlions.

ANTLION LACEWINGS (Family MYRMELEONTIDAE)

In their adult form these lacewings have long, narrow wings which are sometimes patterned. Eggs are laid in dry soil from which the antlion larvae emerge (see p. 79).

Adult antlion lacewing.

DRAGONFLY LACEWINGS (Family ASCALAPHIDAE)

With a wing-span of 70mm these are most conspicuous when at rest as their elongated abdomens are held high above the wings. The body of a male sometimes has a strange protuberance just behind the thorax but they lack the greatly enlarged thorax of a dragonfly. They are found mainly in the drier areas of Australia, particularly in the north-west.

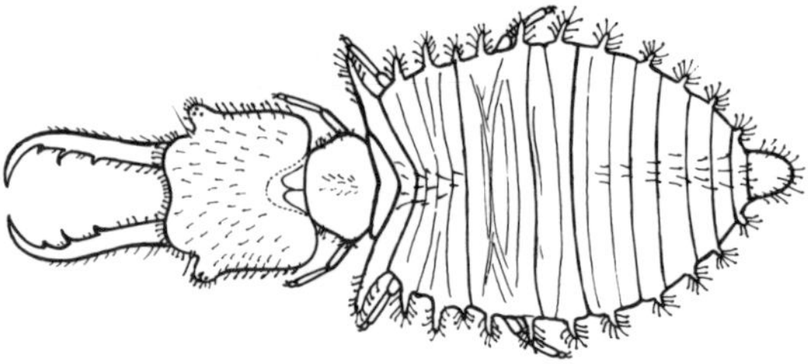

Larva of dragonfly lacewing.

THREAD or **SPOON-WINGED LACEWINGS** (Family NEMOPTERIDAE)

Identified by their extremely long and narrow hind-wings, which are like threads compared with the normal forewings. Larvae of some species have a strange appearance. The neck may be extremely long and there is a conspicuous pincered head. The spoon-winged lacewing (subfamily Nemopterinae) of south-west Western Australia has wings which are narrow at the base and expand into a spoon shape near the apex.

Spoon-winged lacewing. Wingspan 35 mm.

ANTLIONS (Family MYRMELEONTIDAE)

The antlion is included in almost every account of unusual insects. From eggs laid singly in dry soil larvae hatch which bear little resemblance to the fragile adult. The head is large and equipped with a pair of powerful jaws which are used to seize victims. The body is elongated, oval and flattened, and has bristles facing forwards. The hind-legs are enlarged with forward directing claws and these allow larvae to move speedily through the sand.

Antlions occur mainly in areas of low rainfall and the larvae are generally found sheltered by overhangs, under logs, or inside caves. Antlions may often live in pits, which they dig by crawling backwards in a small circle and flipping the sand outwards beyond the edge. When complete the pit is about 25mm to 50mm deep, depending on how steep the sides can be made before the sand will fall. At the bottom of the pit, the antlion lies in wait, buried with only its jaws exposed.

Any insect moving within the edge of the pit starts an avalanche of sand. The antlion reacts by tossing sand up from its lair at the intruder. The flying sand has the dual effect of confusing the intruder and causing a greater avalanche which often causes the insect to tumble to the bottom of the pit. Once there it is grabbed by the sharply-pointed incurving jaws and the antlion then sucks the juice from the intruder's body. When drained, the remains are thrown out of the pit and the trap is again prepared.

Although not too easily obtained, antlions have been captured and raised in captivity in a shallow dish of sand. Food is provided by procuring live ants and other insects which are placed near the edge of the pit and enticed near it with sugar or other bait.

When the larvae are fully grown, after three or four moults, silk, said to be secreted in the Malpighian tubes, is spun through the anus and used to make a cocoon. Within the cocoon, the larva pupates in the soil but works its way to the surface before emerging.

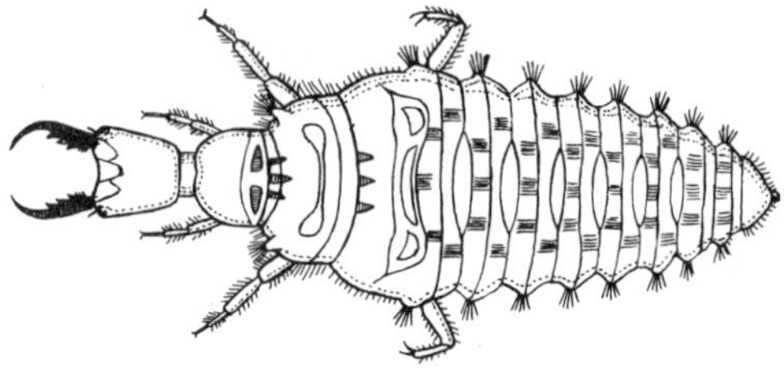

Larva of *Acanthaclisis fundata*, a typical antlion.

APHID-LIONS (Family CHRYSOPIDAE)

The larvae of green lacewings, some living solely on aphids (see pp. 59–60). As each victim is sucked dry, its skin is lifted onto the back of the larva where it remains lodged among stiff, hooked hairs. This load of victims' remains serves as a camouflage against possible predators.

BEETLES

Order COLEOPTERA (sheath-winged)

There are more species of insects than in any other single group of animals and there are more species of beetles than those for any other order of insects. The number of beetle species in the world may be as many as 300 000 (about 40 per cent of all described insect species). The pattern in Australia mirrors world trends and with 19 000 species of beetles described, this figure is nearly double the order with the next most numerous species, the Lepidoptera.

At present, though the classification is still subject to dispute, about 112 families of beetles are represented in Australia and they are grouped in four suborders. For obvious reasons of space, representation here has had to be limited and selection has been made of the most common, the most interesting and those which are of economic importance.

Beetles are easily recognised by their forewings which have been modified and hardened into horny sheath-like wing covers *(elytra)*. The hind-wings are transparent and fold beneath the elytra when not in use. However, with so large a group, there are many variations. The inner margins of elytra commonly meet in a straight line and extend as a roof to the end of the abdomen. However, they may be locked into a single plate; or may be greatly reduced or absent as in rove beetles.

In flight the elytra are held at an angle above the body while the membraneous and wide hind-wings vibrate rapidly.

Beetles are adapted to every form of habitat and the legs vary greatly in form, being adapted in different cases for running, walking, clinging, digging, swimming and for grasping prey.

A beetle's head is hard with two compound eyes and often distinctive antennae. Antennae shapes may be short and thick, bristled, long and thread-like, toothed like a comb, feathery, clubbed or flattened. The mouth is adapted to chewing and grinding and may be used to consume almost all types of material. Ground beetles (Carabidae) will kill a defenceless caterpillar although most other beetles are confined to a diet of plant material. Weevils have the mouth parts located in the tip of a long snout *(rostrum)* which enables them to nibble plants and to bore into fruits and seeds. Longicorns have long mandibles which they use to gnaw their way out of the pupal cocoon and stag beetles have jaws (mandibles) which are very large and project far in front of the head like antlers, though they are used for fighting, not feeding. Depending on the group, beetles may eat live insects or dead and decaying plant and animal matter; they can attack processed materials such as foodstuffs, seasoned wood, furs and museum specimens.

The thorax is divided into three parts and is hard and horny. The upper part between the junction of the head and the base of the elytra is called the *pronotum*. From beneath, the greater part of the thorax can be seen as a large broad structure.

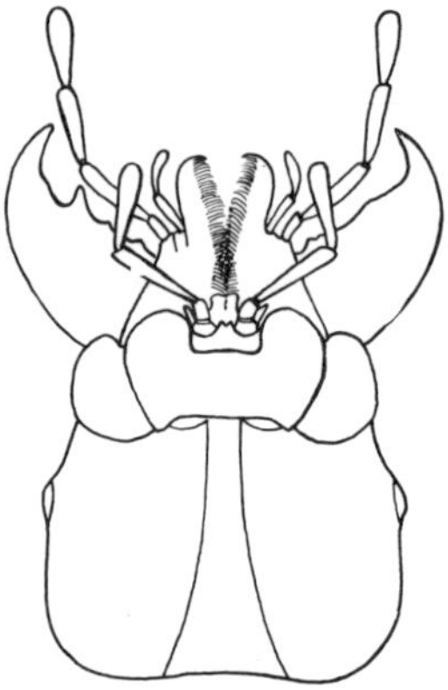

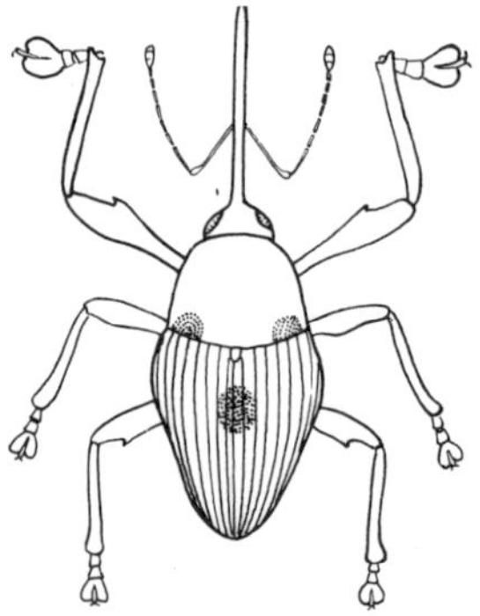

Left: Enlargement of mouthparts of carabid beetle.
Right: Weevil of Curculionidae family.

All beetles pass through a complete metamorphosis. Eggs are often laid near the food which the larvae (grubs) eat and larvae may be provided with long legs and jaws which allow them to pursue and capture insects and other prey. Others may have short legs and fat cylindrical bodies; or have a curved body and be legless grubs.

Duration of larval life varies from a few weeks to several years. The pupa has a hard skin or sheath covering the developing body while the legs and wings remain free. On Lepidoptera, pupal wings and legs are glued down. The pupa may be naked, enveloped in a larval skin, or enclosed in some form of cocoon: and it may be located on a plant, under the bark of a tree, beneath the ground or at the bottom of a pond.

On emergence the beetle's skin is soft and almost colourless but exposure to air hardens the cuticle which darkens or assumes the metallic shades which are typical of some species.

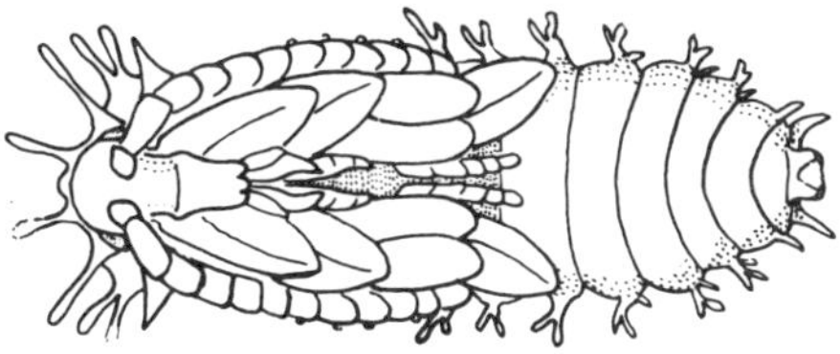

Pupa from Lampyridae family.

In size, beetles vary from microscopic fungus beetles, small enough to scuttle easily through the eye of a needle, to the huge Hercules beetles in the genus *Dynastes* of South America, which may attain a length of 160mm and are larger than many small mammals. The bulkiest beetles are the African Goliath beetles (*Goliathus* spp.) and the South American elephant beetles which have a horn-to-tail length of 115mm. Australia's largest beetle is the carabid *Hyperion schroetteri* found in hollow trees. This species has enormous development of the head and pronotum and may be found up to about 80mm long. The buprestid *Stigmodera grandis* can attain a length of 60mm or more. However, in New Guinea, the native rhinoceros beetles *(Scapanes australis)* which are pests of palm trees, and the longicorn beetles, are much larger.

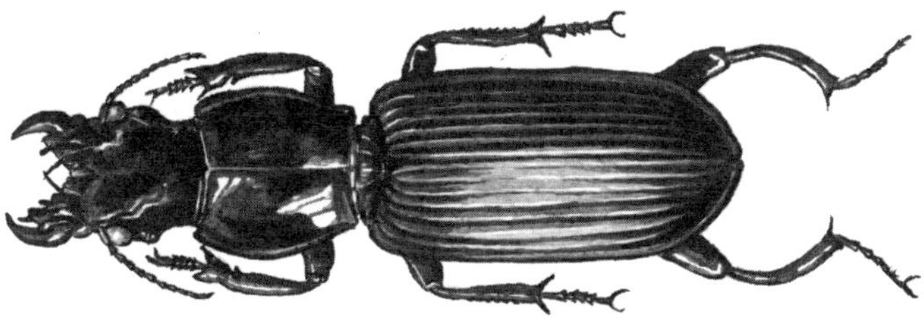

Hyperion schroetteri.

Giant longicorns of other places may grow to a length of 125mm and to a width of 36mm. Others may have antennae far longer than the body.

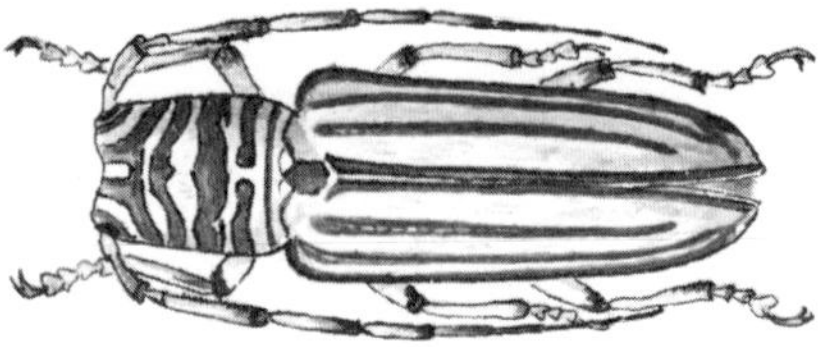

Longicorn beetle, *Rhytiphora dallasi*. Length 40 mm.

Australia's largest longicorn *Batocera boisduvali* may grow to 60mm body length but the distance between the tips of its long antennae when spread at right angles to the body can be 130mm or more.

Some of the earliest known fossil beetles have been found in Permian rocks in New South Wales (some 270 million years old) and also in the USSR. Many families present today had evolved by the early Mesozoic (about 255 million years ago) and they have exploited every possible habitat except the sea.

There are many characteristics used to identify various beetles and it would be impossible here to even attempt to provide a guide to the identification of every beetle family found in Australia.

Groups containing some of the main families may be identified provisionally from the following table. This guide however, is presented only as an aid to identification and does not indicate the true relationship of families to each other as viewed by modern authorities.

Guide to general identification of the most common beetles

See page	*Common name*	*Family*	*Feet (tarsi)*	*Antennae*	*Other*
228	Water tiger beetle	DYSTICIDAE	With 5 distinct segments	Threadlike *(filiform)*	
81	Ground beetles	CARABIDAE			
88–9	Tiger beetles	Cicindelinae (subfamily)			
228	Whirligig beetles	GYRINIDAE			
228–9	Water scavengers	HYDROPHILIDAE	Segments variable in number	Ends broader, often club shaped *(capitate)*	
91–2	Carrion or burying beetles	SILPHIDAE			
92–3	Rove beetles	STAPHYLINIDAE			
110	Sap-feeding or fruit beetles	NITIDULIDAE			
90	Hister beetles	HISTERIDAE			
112–13	Ladybirds	COCCINELLIDAE			
111–12	Handsome fungus beetles	ENDOMYCHIDAE			
111–12	Fungus beetles	EROTYLIDAE			
111	Stem borers	LANGURIIDAE			
106	Dermestids or carpet beetles	DERMESTIDAE			
102–3	Click beetles	ELATERIDAE	Segments variable in number	Third and outer segments similar to teeth on a saw *(serrate)*	
101–2	Jewel beetles or metallic wood borers	BUPRESTIDAE			
103–4	Fireflies	LAMPYRIDAE			
105	Soldier beetles	CANTHARIDAE		Threadlike *(filiform)*	

See page	*Common name*	*Family*	*Feet (tarsi)*	*Antennae*	*Other*
113–5	Pie-dish or darkling beetles	TENEBRIONIDAE	5 segments on front and middle, 4 segments—on back tarsi		
115	Pin-tail beetles	MORDELLIDAE			
93–5	Stag beetles	LUCANIDAE	5 segments on all tarsi	Leaf-like *(lamellate)*	
95–6	Passalid beetles	PASSALIDAE			
96–7	Dung or dor beetles	GEOTRUPIDAE			
97–100	Scarabs or cockchafers	SCARABAEIDAE			
116–8	Longicorn beetles	CERAMBYCIDAE	Tarsi apparently in 4 segments but really 5; 4th hardly visible, 3rd segments have hair on undersurfaces		
119	Pea pests	BRUCHIDAE			
119–22	Leaf beetles	CHRYSOMELIDAE			
123	Primitive weevils	BRENTHIDAE	Tarsi have 4 segments, 3rd usually broad and hairy underneath		Head more or less extended to form snout
122–3	Fungus weevils	ANTHRIBIDAE			
124–7	True weevils	CURCULIONIDAE			
127	Bark and ambrosia beetles	Scolytinae (subfamily)			
102	Feather-horned beetles	RHIPICERIDAE	5 segments on all tarsi	Feathery or fan-like *(pectinate)*	
109–10	Clerid beetles	CLERIDAE		Saw-toothed or clubbed	

GROUND BEETLES (Family CARABIDAE)

With 1600 described species in Australia, the carab or ground beetles are one of the largest families present. There is much variety in their appearance and the various groups have been classed ecologically into three distinct types. They are: ground dwellers which do not live near water; ground dwellers which live close to water; tree dwellers.

As a family, the beetles usually are medium sized and dark coloured. Many may be found under rocks, stones or bark where they prey on other small animals. Their antennae usually have 11 segments and are either thread-like or flattened. Wings are sometimes absent.

Both adults and larvae are active and fast runners, and the larvae are almost always similar to caterpillars *(campodeiform)* with a 10-segmented abdomen, and legs which each have two claws.

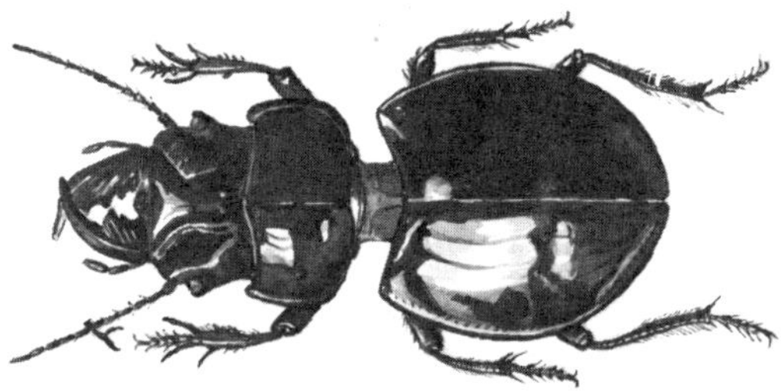

Carab beetle, *Euryscaphus waterhousei*. Length 60 mm.

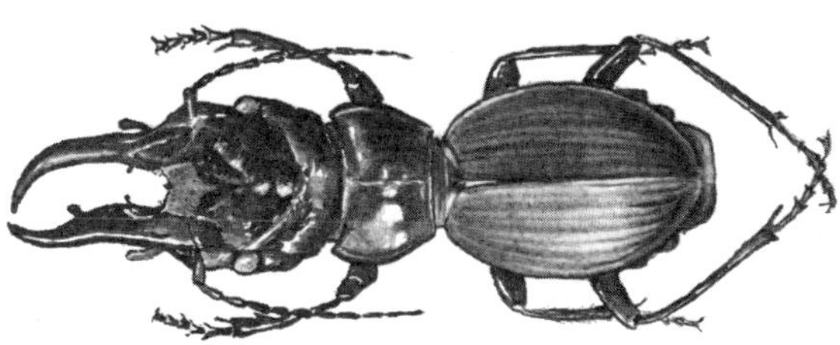

Carab beetle, *Mecynognathus dameli*. Length 55 mm.

Many ground beetles serve a useful purpose, destroying pests of many kinds. *Calosoma schayeri*, particularly, which preys on army and cutworms (Lepidoptera) tearing them to pieces with its powerful jaws.

In the subfamily Paussinae, one of the strangest carabs is the species *Arthropterus planicollis*, which may be identified by its greatly expanded paddle-shaped antennae, whose function is not known. These are small beetles, from 5 to 13mm long and dark brown. Many live with ants although adults of the Australian species do not have abdominal glands which emit secretions on which the ants live. It is possible these may be present in the larvae.

TIGER BEETLES (Subfamily CICINDELINAE)

Conspicuous for the head, which is turned down vertically and contains powerful projecting and curved jaws, many species in this subfamily are brightly coloured or metallic-looking and they actively chase and capture their prey by their superior speed. Some tiger beetles are associated with salt water, either on the banks of inland saline lakes or on the seashore. They may also be found on sandy heaths, and sand-banks. Some species live in trees. Tiger beetle larvae live in burrows in the sand and often rest with the heavily armoured head level with the surface. The larva grips the wall of the burrow with two recurved hooks on the top of the fifth abdominal segment and when an insect passes close enough, it is seized. Should the prey be large, the hooks prevent the larva being pulled from its burrow.

Some 65 species of tiger beetles have been described in Australia and the majority are known from the north of the continent. Only three species occur in the east and south. The most numerous species are in the genus *Cicindela* which is bronze green with the elytra margins outlined in white. This is often found on the banks of creeks and dams in inland areas. On the beaches of New South Wales and Victoria the cream coloured *Cicindela ypsilon* has coppery markings resembling the Greek letter 'y'. Its colour blends with the sandy background although it is a fast runner and readily takes to the air.

The striking green or purple large-headed tiger beetles (*Megacephala* spp.) are found from Queensland to South Australia in inland areas. They are nocturnal and may not fly.

Tiger beetle of the *Cicindela* genus. Length 20 mm.

BOMBARDIER BEETLES (Subfamily BRACHININAE)

Although some 500 are known, only one bombardier beetle is present in Australia. It is often found in the damp earth beneath rocks. Its black elytra are marked with four yellow spots. Its main claim to fame however is based on its unusual defensive mechanism. When the bombardier beetle is disturbed, it will be seen to release, with a small explosive sound, a vaporous cloud from glands at the end of the abdomen. Investigations have shown that the secretion contains various chemicals including hydrogen peroxide and the 'explosion' occurs inside a specially stiffened sac within the beetle. The proportion of hydrogen peroxide (28 per cent) found in the bombardier beetle's 'gas' is the highest known in any living organism.

Largely restricted to Australia, the subfamily Pseudomorphinae are a widespread group which live under tree bark. The larvae of one species prey on meat ants (*Iridomyrmex purpureus*), digging their burrows in the vicinity of the ant beds.

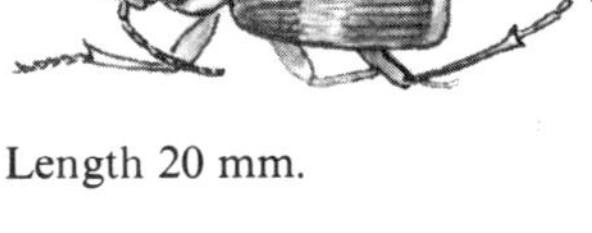

Bombardier beetle. Length 20 mm.

CARRION or **HISTER BEETLES** (Family HISTERIDAE)

Rarely larger than 18mm long, carrion beetles are usually black with metallic tints and have elbowed antennae with a hairy three-segmented club and prominent jaws. Some of these beetles remain flattened for a life under tree bark but most are hemispherical. They live on a variety of decaying organic compounds including decaying animal and vegetable matter, and dung. Some may even live on blowfly maggots.

When disturbed, a carrion beetle will roll on its back, pulling its legs close to the abdomen, or will remain motionless, giving the appearance of a small black seed.

Some carrion beetles have distinct grooves at the base of the elytra and these may be useful in species identification. The elytra never extend to the full length of the abdomen.

One of the widest known of Australia's 144 described species is *Hololepta sidnensis*, which is black with projecting jaws and is the largest known in this country. A smaller metallic blue-green carrion beetle, *Saprinus cyaneus* is also common. Species in the genus *Chlamydopsis* have bristle-covered elytra and live in ants' nests; others may spend their lives in termite mounds, or in the nests of certain birds or mammals.

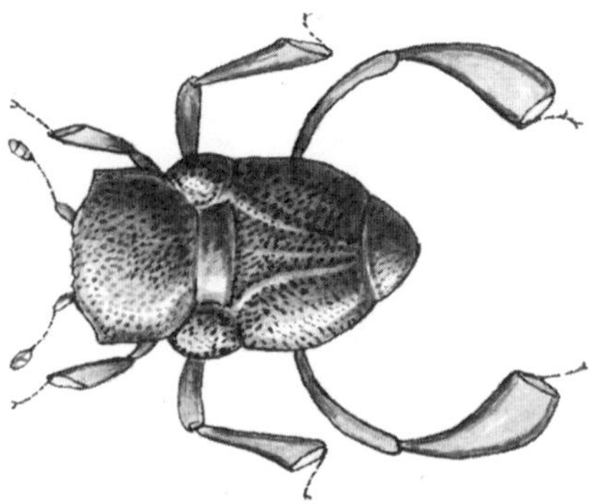

Hister beetle of the *Chlamydopsis* genus. Length 4.5 mm.

BURYING BEETLES (Family SILPHIDAE)

Australia has only three species of burying beetles, but their sizes may range up to 40mm long. Two species are found mainly in the south-east of the continent and in Tasmania (*Ptomaphila lacrymosa* and *P. perlata*). In New Guinea and Queensland, the species *Diamesus osculans* continues a distribution which extends to India.

These beetles are flattened with clubbed antennae. The southern species have elytra which completely cover the abdomen and are dull yellow-brown with elongated raised markings. The middle of the elytra is dark whilst the edges are paler.

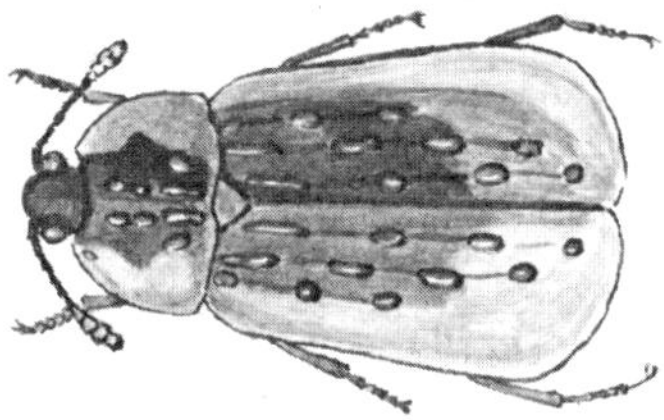

Ptomaphila lacrymosa. Length 30 mm.

To date, little investigation has been devoted to Australian burying beetles, but overseas the 200 species are famed for their odd behaviour. Burying beetles live exclusively on carrion, a dead animal (mouse, rat or bird) often being spotted first by a male scout. The male releases an attractant scent and calls a female by using its stridulatory mechanism. When another beetle arrives, both begin to dig away the earth from beneath the dead animal until, in due course, the animal subsides into the soil, and is completely covered. A mouse can be buried in two to three hours and a rat in five to six hours.

In overseas species, the male makes a burrow from the buried food supply to the surface and lines it with fur or feathers from the buried creature. The female then makes a nesting tunnel leading away from the cell containing the carrion. At the end of the nesting tunnel about 15 eggs are laid in a chamber and the female then returns to the carrion to feed. When the young hatch, they join the female at the carrion. They do not eat the carrion, but are fed predigested food by the female. The larvae moult twice in four or five days, and until the second moult is completed, they are unable to feed themselves. Some burying beetles live on snails and vegetation.

ROVE BEETLES (Family STAPHYLINIDAE)

Superficially resembling earwigs, although lacking the abdominal 'pincers', rove beetles have slender, straight-sided bodies and elytra which are shortened so that at their longest at least three abdominal segments remain uncovered. At their shortest, only two are covered.

The abdomen is soft. When the beetle is running the abdomen is raised as though it might be used as a sting. However, there is no sting, although nearly all species are able to eject an anal secretion. In most species this secretion is harmless to man, but in species of *Paederus*, including the Australian *P. cruenticollis*, the fluid may cause severe skin blistering which has gained the species the name of the 'whiplash rove beetle'. This linear mark, often caused by brushing away the beetle, appears a day or two after contact and frequently the cause is forgotten by the time the injury is manifested. This species is usually found on the banks of rivers. It is very slender with a steel-blue body and a red band across the abdomen.

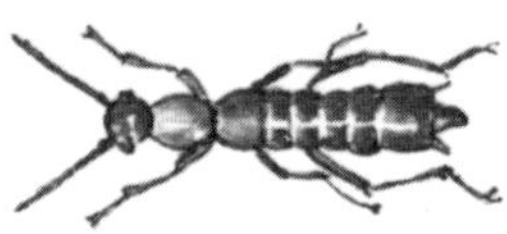

Rove beetle of *Paederus* genus. Length 12 mm.

The commonest and largest species known in Australia is sometimes called 'the devil's coach horse' *(Creophilus erythrocephalus)* and is about 20mm long. It has a bright orange-red head with a central black spot and a black body, antennae and jaws. If disturbed, this rove beetle faces the threat, raises its head and turns up the tip of its abdomen. Despite the shortness of the elytra, the hind-wings are of normal length and overseas reports suggest that both the upturned abdomen and the hind-legs are used to fold the hind-wings beneath the small elytra.

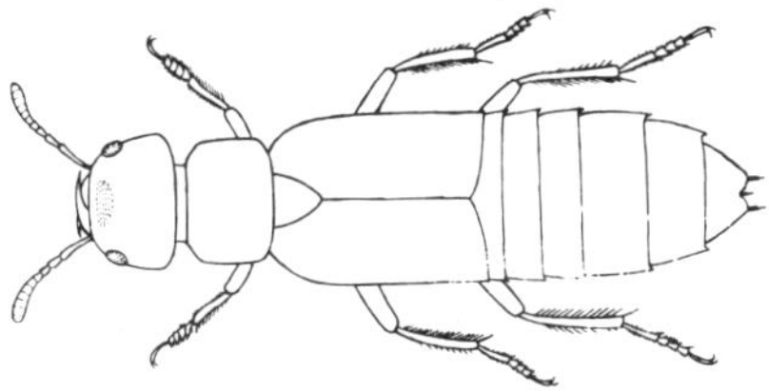

Creophilus erythrocephalus. Length 15 mm.

In Australia 650 species of rove beetles have been described, compared with 27 000 known in the world to date. Many are found in decaying material and in damp locations. The 10mm long, shining black species *(Cafius sabulosus)* lives on beaches, where large numbers assemble under seaweed stranded on the shore. Some smaller species of rove beetles live in ant and termite nests. Some may be scavengers of ant or termite debris while others may prey on eggs. The Australian species *Myotyphlus jansoni* is wingless and is found in the fur of native rats.

Rove beetles in the genus *Stenus* have large round eyes and the antennae are located between them. They can skim across the surface of water and generally live in marshes and beside streams. They employ an anal secretion which lowers the surface tension of the water behind them as they skim and this actually 'pushes' them along. Bulging eyes are also found in species of the genus *Megalopsidia*.

STAG BEETLES (Family LUCANIDAE)

Well-known for their conspicuous antler-shaped jaws, the stag beetles are striking insects with impressive metallic colours. Their adult length reaches 50mm. The jaws *(mandibles)* of males are often much larger than those of the females and their function is considered to be for fighting with other males during natural and sexual selection. Sometimes the inner edge of the jaws is toothed but on other species they may be covered with a form of fur-like bristles. Antennae are clubbed, though sometimes they resemble a comb. The thorax is as wide as the abdomen and elytra and there is a narrowed waist between the two parts of the body.

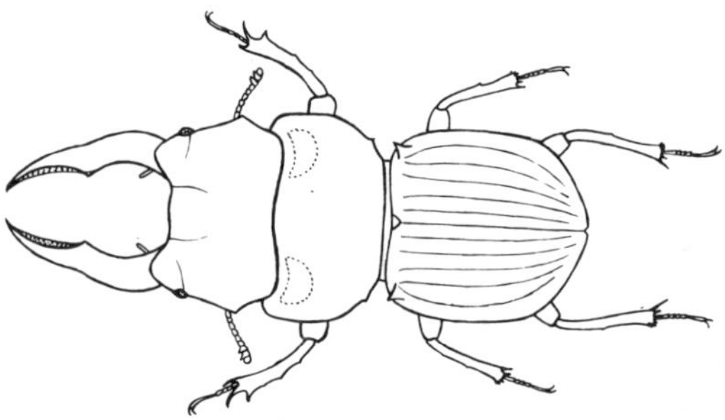

Stag beetle, *Hoplogonus simsoni*. Length 30 mm.

Eggs are laid in rotting logs and larvae, which are fat white grubs, feed in the wood and eventually pupate in oval cells formed of wood fragments. In some places, the life-cycle lasts several years.

Of the 750 described species of stag beetles in the world, the Indonesian species *Odontolabis alces* is the largest with a length of 100mm.

The largest of Australia's 75 described species is *Phalacrognathus muelleri*, a large bronze-green and copper-tinted stag beetle which is found in tropical north Queensland rain forests and is uncommon. Most species appear to be confined to tropical and sub-tropical areas but some are found in Tasmania and Western Australia.

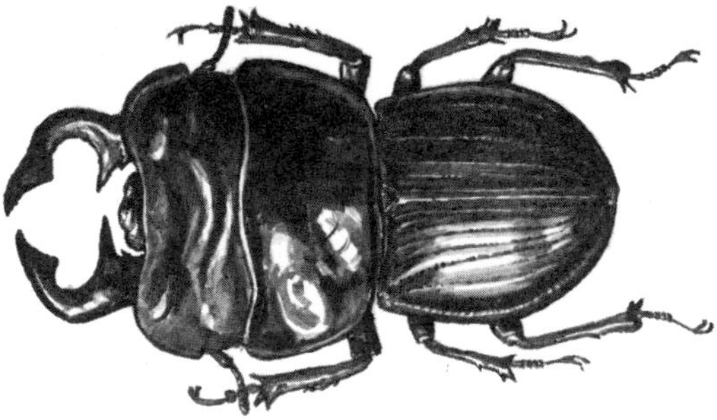

Stag beetle, *Lissotes acmenus*. Length 40 mm.

The king or gold stag beetles, *Lamprima latreillei*, are often found on gum trees in eastern Australia during November and December where they feed on young shoots. A related species *L. mandibularis* has jaws which are long and upswept. Colours in this genus range from gold to greens, blues and purples. The golden stag beetle *(Lamprima aurata)* is found in casuarinas beside New South Wales rivers.

PASSALID or **'BESS' BEETLES** (Family PASSALIDAE)

Both adults and larvae possess distinctive stridulatory mechanisms and the passalids are known in the USA as 'bess' beetles after the 'bess, bess, bess' sounds which they make. With adults, this creaking or hissing is made by vibrating the abdomen on the elytra. With the soft, greyish white grubs, the hind-legs are reduced to clubs and these stridulate on a file on each centre leg.

Adult passalids are large, glossy black beetles with the pronotum as wide or wider than the abdomen, and with a narrow waist between the two main body parts. Elytra are deeply grooved and the antennae are curved and clubbed.

Only 34 of the world's 490 described passalid species have been recorded from Australia and these mostly inhabit the tropics of Queensland and New Guinea. All live under decaying logs and if Australian species are similar in behaviour to those elsewhere, they are highly social insects.

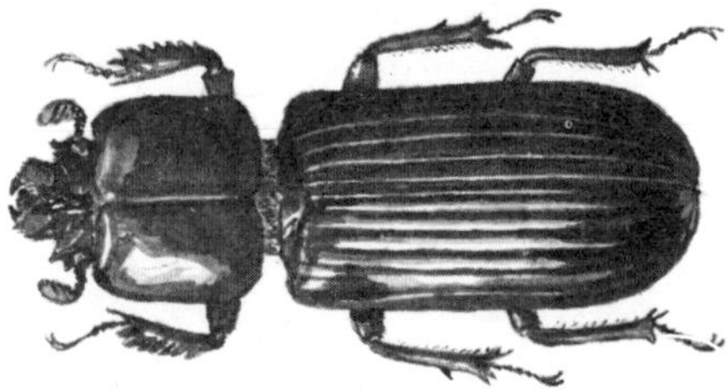

Passalid of the *Aulacocyclus* genus. Length 45 mm.

Females lay their eggs in burrows and after the young hatch, they are fed by the females with masticated wood.

When disturbed, adults and young both stridulate, possibly as a warning; on other occasions, stridulation may be a means of social communication.

Only one species *(Pharochilus politus)* is known from Tasmania. The most common species in south-eastern Australia are *Pharochilus dilatus* and *Aulacocyclus edentulus*. Overseas species attain adult length of 75mm but Australia's largest species, in the genus *Mastochilus*, found in Queensland's tropical rain forests, does not attain a length of more than 50mm.

DUNG BEETLES (Family GEOTRUPIDAE)

Possessing a high rounded body, often surmounted by a striking 'horn' on the male, these 7 to 25mm long beetles usually live in well grazed paddocks where they dig nesting burrows alongside cattle droppings. They collect the dung and move it to the burrows where the female lays her eggs. (This activity may well be useful in fertilising soil where numbers of nesting beetles are large.)

Dung beetles are able to stridulate by rubbing a hind-leg on the abdomen. They have 11-segmented clubbed antennae unlike Scarabaeidae (see below) which have only nine to 10 segments. To date 81 species of dung beetles have been described in Australia and most are included in the genus *Blackburnium* (previously *Bolboceros*).

The species *Blackburnium reichei* is large and impressive with its two strongly-divided horns on the pronotum and a slender curved horn on the head. The Tasmanian *Elephastomus proboscideus* has a slender protuberance projecting from the head, bearing some resemblance to an elephant's raised trunk.

SCARABS (Family SCARABAEIDAE)

With more than 17 000 species throughout the world and some 2100 scarabs described from Australia, this family is one of the largest among beetles. All have stout bodies. The head is never bent down and the antennae never have more than 10, and often less segments, flattened at the tops into discs *(lamellae)* or leaves. The larvae are curl-grubs with a pale reddish-brown head and three pairs of legs, with the hind pair usually longer than the others. Front legs may be adapted for digging *(fossorial)*. The latest classification lists nine subfamilies in the Scarabaeidae of Australia but only four are described below. They are:

	Subfamily
Dung-ball rollers	Scarabaeinae (Coprinae)
Cockchafers	Melonthinae
Christmas or king beetles	Rutelinae
Goliath beetles	Dynastinae

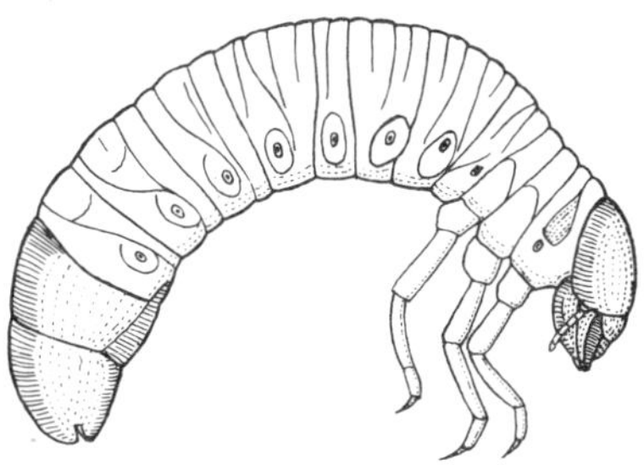

Scarab larva.

Dung-ball rollers

These are the famous scarabs of ancient Egypt, whose image was so often included on carvings made from the times of the first Pharaohs. It has been assumed that the beneficial behaviour of these beetles was the reason for their being so immortalised. Like the dung beetles (see above), the dung-ball rollers live on animal refuse and pack this into burrows in the soil where it is slowly eaten, thus assisting in the fertilisation of the soil. However the dung-ball rollers have broad front legs which they use both for handling the dung and digging their burrows. When the beetle first obtains the dung, it rolls it into a perfect sphere and then rolls it into cavities in the soil or under stones. It has been estimated that after obtaining this food, which is then stored underground, a single beetle can eat more than its own weight in 12 hours. Similar balls are also made in which the female lays her eggs. After hatching, the larva eats the inside of the pellet, and also pupates within it.

Australian dung-ball rollers range in adult length from 2 to 25mm. The body is broadly oval and usually black. The last abdominal segment *(pygidium)* is never covered by the wing covers *(elytra)* and there is only one spur on the hindmost leg *(tiba)*.

The Australian species *Macropocopris symbioticus* spends the whole of its adult life clinging to the fur around the anus of wallabies and feeding on excreta.

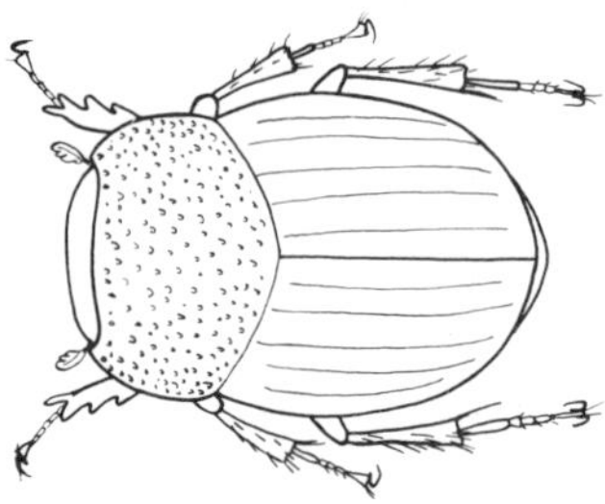

Macropocopris symbioticus.

Cockchafers

These form the largest subfamily, with 1400 species described in Australia. Length ranges from 5 to 40mm and they are usually black or reddish-brown.

Eggs of all species develop into curl-grubs which hatch below the surface of the soil. They live on plant roots and at times can become a pest. The introduced grey-backed cane cockchafer *(Dermolepida albohirtum)* is a significant pest of sugar cane while 'white grubs' often cause damage to vegetables and wheat. Adults are leaf or nectar eaters and in some years plague numbers of the small metallic green or purple scarab (genus *Diphucephala*) may swarm over the leaves of native *Eucalyptus* and wattle trees and completely defoliate them. Other species of cockchafers may be found in termite mounds.

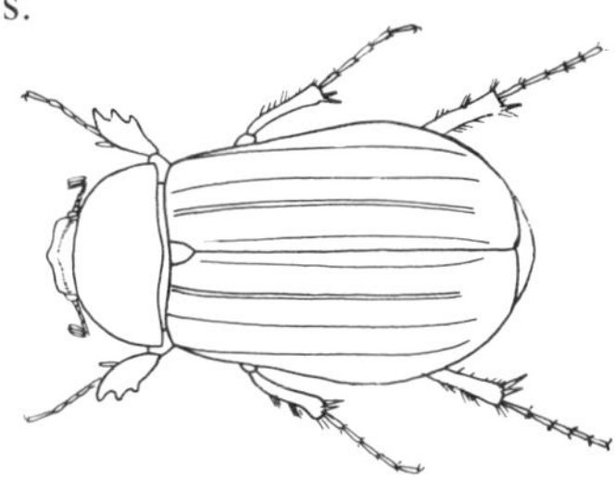

Cockchafer beetle. Length 10 mm.

Christmas or **king beetles**

These are perhaps most noticeable in summer, when they may be attracted by lights, often falling to the ground, temporarily dazed. They are broad-bodied and may range in length from about 8 to 25mm. Metallic-coloured golden or green beetles in the genus *Anoplognathus* have large, unequal and movable claws. The larvae, as with other scarabs, live underground and may damage lawns and pastures. Usually these beetles are found in the higher rainfall areas. In Western Australia, the best known Christmas beetle is the grey-green species *Anoplostethus opalinus* which also has a purple opalescence.

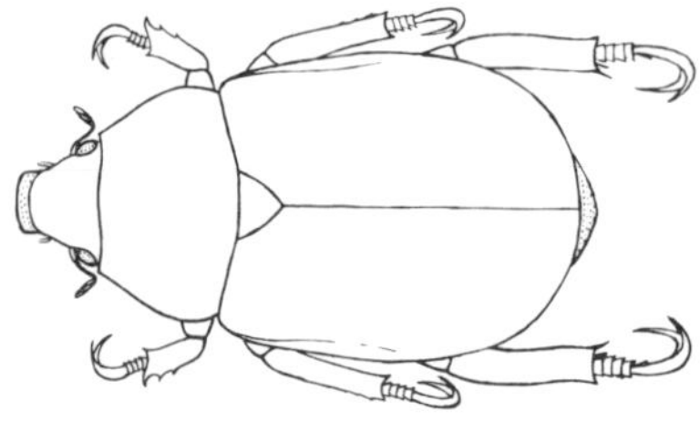

Christmas beetle of the *Anoplognathus* genus. Length 35 mm.

Goliath or **rhinoceros beetles**

Sometimes attaining a length of 70mm, these large beetles have heavy bodies and claws. Males often have large horns projecting from the head. Possibly the most impressive Australian species is the dark brown *Xylotrupes gideon* whose male can develop two horns on the pronotum and a stout curved horn on its head. The females, however, have no conspicuous horns and the thorax is round like other scarabs. When picked up they make a hissing sound. Reports indicate that while grubs live on roots and decayed vegetable matter in the soil, adults may eat bark. The introduced South African black beetle *(Heteronychus arator)*, is a pest of grain and grass in New South Wales.

Goliath beetles in the genus *Cryptodus* are found with ants and adults have the antennae and mouth protected by extensions of the chin *(mentum)*.

Rhinoceros beetle of the *Xylotrupes* genus. Length 60 mm.

JEWEL BEETLES (Family BUPRESTIDAE)

These beetles, as their name suggests, contain many species famed for their attractive metallic tints and yellow, orange, red and blue colour patterns. In many places, jewel beetles have been used to make jewellery and were worn unmounted in the hair and on garments.

About 800 species have been described from Australia and some grow to a length of 60mm. Most are elongated, with thick outer surfaces *(cuticle)*. Their antennae have 11 segments and tarsi on all legs are five-segmented.

The larvae are wood borers, and tunnel in the trunks or roots of living trees. Some species seem particularly adapted to certain host trees and from such associations some jewel beetles have been given specific common names. These include the cypress-pine jewel beetle *(Diadoxus erythrurus)*, whose adult is brightly coloured in green, brown and yellow. There is also the hoop-pine jewel beetle *(Prospheres aurantiopictus)* from Queensland, which is shining black with eight yellow marks on the elytra.

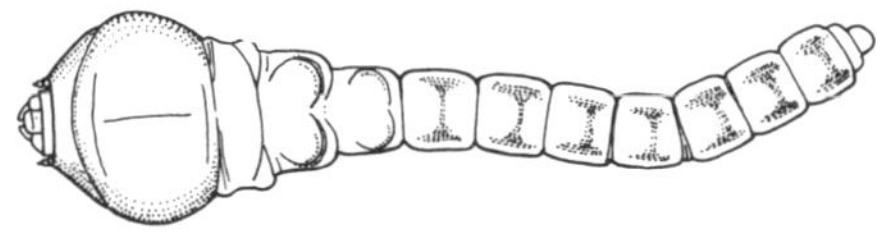

Stigmodera larva. Length 35 mm.

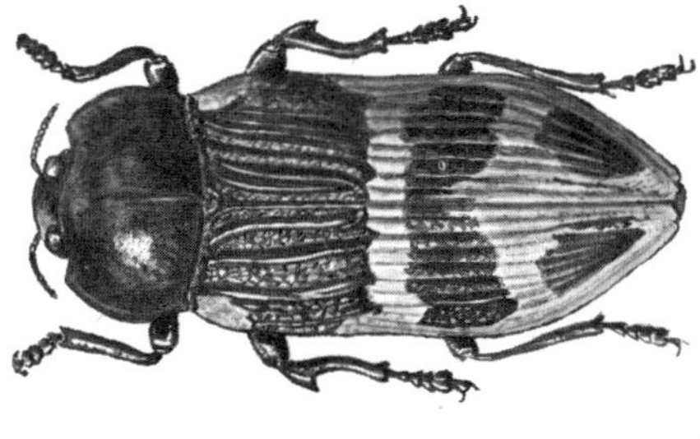

Jewel beetle of *Stigmodera* genus. Length 60 mm.

About half the Australian species are included in the genus *Stigmodera* and most have striking colour combinations of red, yellow and black iridescence. The largest Australian jewel beetle is the Western Australian *Stigmodera heros* which grows to a length of 75mm and is chestnut brown. Some of the best known of these beetles are in the species *Merimna atrata*. Known as 'camp-fire beetles', they are black and rather flat, with a length of 15 to 30mm and tend to fly into camp fires.

FEATHER- or **FAN-HORNED BEETLES** (Family RHIPICERIDAE)

The most conspicuous feature of these elongated beetles, which grow to a length of 25mm, is the fan-like antennae on the males. Females have toothed, rather saw-shaped antennae. The northern New South Wales species *Rhipicera femorata* is black with spotted elytra. To date, 11 species of feather-horned beetles have been described from various parts of Australia including Mornington and Echuca in Victoria.

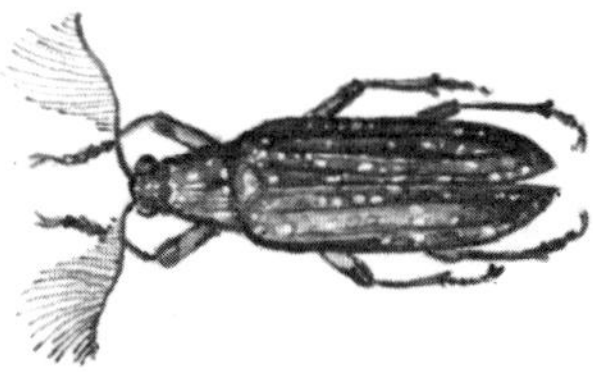

Rhipicera femorata. Length 18 mm.

CLICK BEETLES (Family ELATERIDAE)

Torpedo-shaped or elongated and sometimes attaining a length of up to 50mm in Australia, these leaf-eating beetles have a unique 'clicking' mechanism. If disturbed, they fall to the ground, often resting on their back. To regain their feet, they have a spine on the underside of the prothorax which slides in a groove on the underside of the abdomen. When lying on its back, the beetle straightens its body,

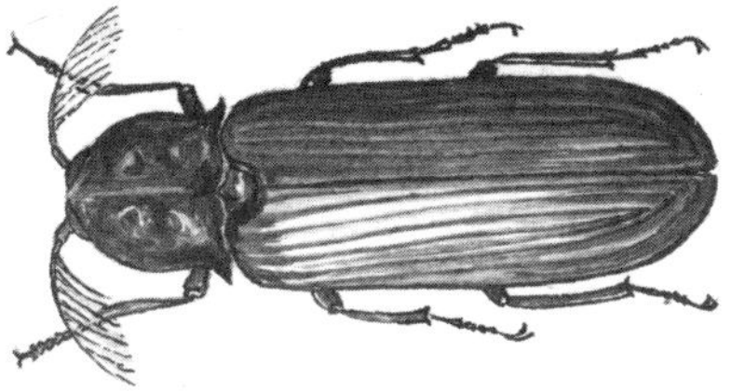

Click beetle, *Tetralobus corrosus*. Length 45 mm.

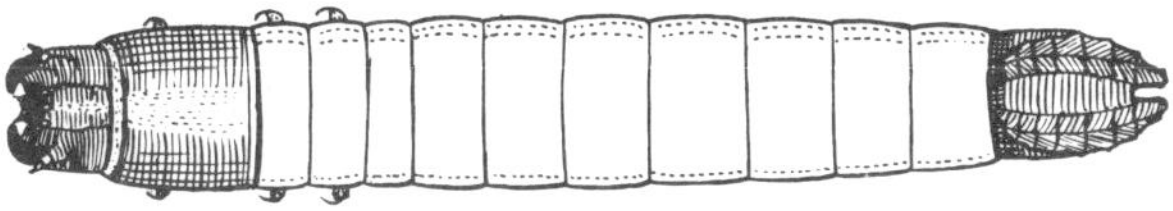

Wireworm, or click beetle larva. Length 20 mm.

bending the thorax farther back until the spine reaches the end of the groove, somewhat like cocking a gun. Suddenly the spine is released, accompanied by a distinct 'click' and the release of muscular tension forces the base of the elytra against the ground with force enough to throw the insect into the air. While in the air, it attempts to turn and land on its feet.

Larvae of the click beetles are the notorious 'wireworms' which are pests of root crops and grasses. They are long, thin, cylindrical and smooth and usually straw-coloured with three pairs of legs. Some wireworms of the genus *Paracalais* prey on larvae of larger wood-boring beetles, especially those of longicorn beetles. Some larvae may grow to a length of 60 to 70mm.

FIREFLIES (Family LAMPYRIDAE)

Confined mainly to the northern parts of Australia and extending south only to the Blue Mountains of New South Wales, fireflies are fascinating to watch. Although there are some 1700 species known in the world, only 16 have been described from Australia.

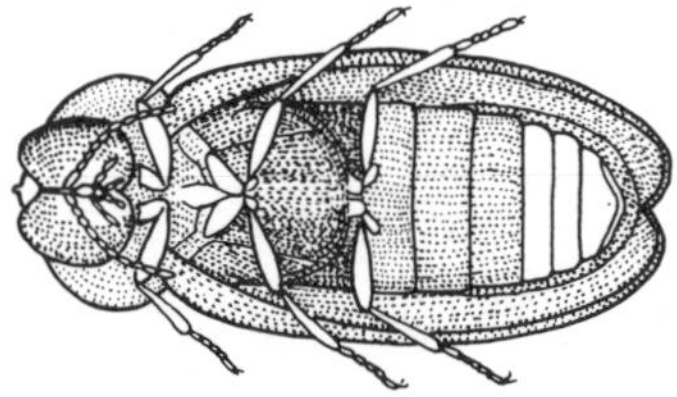

Ventral view of firefly of *Atyphella* genus showing light producing area in white.

Eggs, larvae and adults all have luminous properties, although the light produced by adults is by far the brightest. The light is the result of oxygen being supplied to an enzyme, luciferase, which is located in cells at the end of the underside of the abdomen. These beetles exert control over their light production and it is thought that the flashing serves as a sex attractant and mating signal. In the Australian genera *Luciola* and *Atyphella*, light is produced by both sexes. Females are winged but may not fly.

Firefly beetles rarely attain lengths of more than 6 to 12mm. They are elongated, with long-toothed antennae, usually with 11 segments. Their elytra are soft and feathery. The Blue Mountains firefly *(Atyphella lychnus)* is light brown with dull yellow stripes on the elytra. The northern Australian and Torres Strait islands firefly *(Luciola costata)* is yellow with black-tipped elytra.

Larvae are long and brown with conspicuous body segments. They are carnivorous and overseas species prey on snails by injecting a digestive enzyme into the snail's body. This liquefies the flesh and it is then sucked from the shell.

Firefly larva.

SOLDIER BEETLES (Family CANTHARIDAE)

Like the fireflies, the soldier beetles have soft elytra and as their name suggests, are carnivorous predators both in their larval and adult stages. Adults are brightly-coloured, rarely more than 18mm long, and have a distinct head. Their claws may be simple or comb-like *(pectinate)*. Some soldier beetles in the genus *Sitarida* are wasp mimics. Others may be yellow or metallic blue-black.

Eggs are laid in the ground but the larvae prey on eggs of locusts and flies. Adults also prey on soft-bodied insects and may often be found on flowers. The species *Chauliognathus pulchellus* is often found in the mountains of New South Wales during the warmer months.

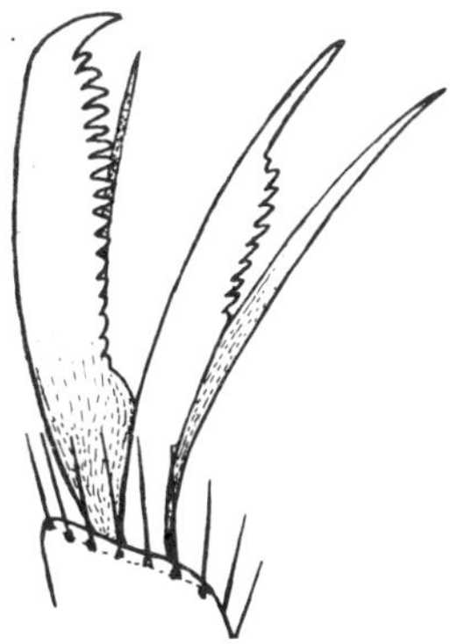

Left: Soldier beetle of *Palaestra* genus.

Right: Soldier beetle claw showing pectinations and elongated processes. Length 20 mm.

FLOWER-DWELLING BEETLES (Family MELYRIDAE)

Similar to soldier beetles with their soft elytra and rather shorter and bristle-covered bodies, the flower beetles are also carnivorous and may be bright yellow and blue. One species, *Laius bellulus*, is a rice pest in New South Wales. The antennae of males of this species have a swollen or distorted third segment.

DERMESTID BEETLES (Family DERMESTIDAE)

One of the most pestilential families of beetles, these insects sometimes as adults and invariably as larvae, cause extensive damage to animal products including furs, wool, feathers, leather and even some processed foods.

They are small and rarely exceed a length of 9mm and their oval, deep shells are covered with dense hairs or scales which may be mottled. Dermestid beetles have clubbed antennae with three to five segments and with the exception of the introduced genus *Dermestes*, possess a simple eye *(ocellus)* between the compound eyes.

To date, 89 species of dermestids have been described in Australia and these include the introduced bacon and larder beetles *(Dermestes vulpinus* and *D. lardarius)* which may be found under dead animals and in bacon and cheese.

Museum beetles (genus *Anthrenus*) are well-known pests of insect collections, mammal and bird skins, and fur coats and woollens placed in store. Adults of this genus are pollen feeders but their hairy larvae with conspicuous clumps of hairs at the tail can cause considerable damage to carpets, blankets and other woollens.

A common native species *Trogoderma froggatti* is small and black and lives under tree bark, feeding on dead insects.

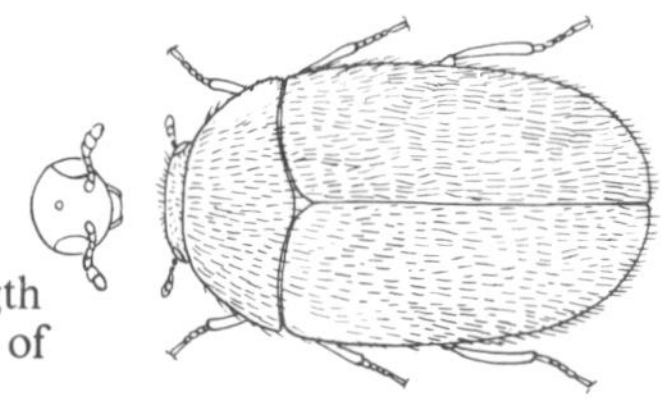

Dermestid of *Trogoderma* genus, length 4 mm. Also, front view of head showing *ocellus*.

One use has been found for the common hide beetle *(Dermestes maculatus)*, in the preparation of specimens for zoological collections. The larva of this beetle is placed in a cage with the body of a small dead vertebrate. The insect will clean the skeleton of all hide and flesh, at the same time leaving the most fragile bones in place.

Furniture borers and other pests

Superfamily BOSTRYCHOIDEA

Under the latest classification, the following families of beetles, which other authors have classed as subfamilies, have been granted family status within the above-named superfamily. All are smallish, with the head often inconspicuous beside the large prothorax. Antennae vary but all these beetles feed on wood, dry vegetables or animal materials.

FURNITURE BORERS (Family ANOBIIDAE)

Rarely larger than 6mm, these brown or red beetles are wood borers and have the head hooded beneath the thorax. The best known species is the introduced furniture beetle *Anobium punctatum*. The beetles deposit eggs on the surface of timber and the small white grubs bore into the wood. Larvae live in well-seasoned timber in houses and in furniture. An overseas species not present in Australia, *Xestobium rufovillosum* is the famed 'death watch beetle'. The biscuit beetle *(Stegobium paniceum)* and the tobacco beetle *(Lasioderma serricorne)* are anobid pests of stored products. Many of these beetles and their larvae have yeasts in the gut which may aid digestion. There are about 100 Australian species.

GOLDEN SPIDER BEETLES (Family PTINIDAE)

Never exceeding 5mm in length, these beetles have a small head and prothorax and a large globular abdomen entirely

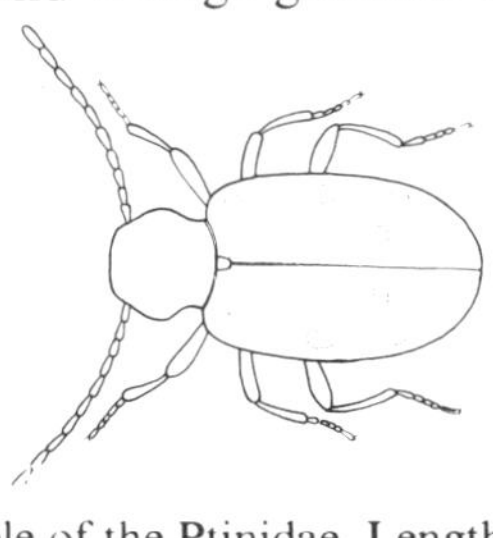

Ptinus exulans, an example of the Ptinidae. Length 3 mm.

covered by the elytra. Their antennae are thin and their legs stout. In the wild, they live on decaying plant and animal matter, but are common in birds' nests and those of other fauna. When access is gained, some species will also damage stored food and skins.

AUGER or **SHOT-HOLE BEETLES** (Family BOSTRYCHIDAE)

Growing to a maximum of 20mm in length, these thick-set cylindrical beetles have a three to four-segmented antennal club. The elytra is usually rough and often spined. The larvae bore in unhealthy trees and drying timber, living on wood starch and sugars and making cleanly-cut tunnels and exit holes. The larvae do not infest seasoned timber but may emerge after earlier ingress.

The largest of Australia's 40 described species is the black and spiny auger beetle *(Bostrychopsis jesuita)* which grows to 20mm in length and is known to feed on gums and acacia trees. The introduced species *Rhyzopertha dominica* is a serious grain pest. Similar beetles of the species *Dinoderus minutus* cause damage to cane furniture and bamboo.

Bostrychopsis jesuita. Length 17 mm.

POWDER-POST BORERS (Family LYCTIDAE)

With five species described from Australia of the 70 known in the world, these small and narrow, parallel-sided beetles attack sapwood in building timber. The larvae feed in dry wood and roots. The most serious pest in the family is the introduced powder-post beetle *(Lyctus brunneus)*. The surface of timber in which larvae of this pest have been active may be left intact, but the interior will have been transformed into powder. The life-cycle lasts about a year and most of the cycle is spent in the larval state.

Superfamily CLEROIDEA

CADELLE BEETLES (Family TROGOSSITIDAE)

There are 27 described species in Australia, all broad and mottled with parallel sides. Antennae are 10 or 11-segmented and clubbed. In some species, the upper surface is covered with bristles or scales (e.g. *Leperina turbata*). The larvae are predatory and live beneath bark, or in tunnels of wood-boring insects. They are generally associated with trees. The introduced cadelle beetle *(Tenebroides mauritanicus)* is a pest of grains, flour and dried fruit but will also attack larvae of other insects which live on these foods. These shiny black and flattened beetles can survive for long periods without food.

CLERID BEETLES (Family CLERIDAE)

Most of Australia's 300 described species of clerid beetles are predatory in both adult and larval stages, the larvae preying on the larvae of wood-boring beetles. Adults may attain a length of 40mm and the largest Australian species, *Eunatilis titana*, is dark brown with deeply grooved elytra.

An attractive species is the 18mm long yellow-horned clerid *(Trogodendron fasciculatum)* which is shiny black with a 'brush' of bristles on each forward edge of the elytra, and a white-bordered velvety area at the back of the elytra. Its name comes from the yellow antennae and the two yellow spots on each elytron. This insect may often be found on tree trunks during the heat of the day. Superficially clerid beetles resemble certain wasps and if handled will bite and retain

Trogodendron fasciculatum. Length 15 mm.

their grip. The larvae attack those of longicorn beetles inside their tunnels. One clerid pest is the red-legged ham beetle (*Necrobia rufipes*) which infests fatty or oily produce such as ham and copra. Many clerids are brightly coloured and patterned.

FRUIT or **SAP-FEEDING BEETLES** (Family NITIDULIDAE)

These are small dark beetles, with the elytra squared at the back and showing some abdomen from above. They may often be found near recently felled trees or fallen ripe fruit, decaying fungi and carrion.

Antennae have 11 segments and the three last segments usually form a club. In Australia, 120 species have been described, a small proportion of the 2200 known throughout the world.

Some fruit beetles in the genus *Carpophilus* are pests of dried fruit and may be dark brown with a yellow patch on the shortened elytra. The Kurrajong pod beetle (*Circopes pilistriatus*) is hairy brown and spends its life-cycle in the seed-pods of the kurrajong tree. Green husks may be opened carefully to reveal many of these beetles which remain there until the husk dries and splits permitting the adults to escape.

Fruit beetles of the genus *Brachypeplus* are found in native bee nests and live on pollen.

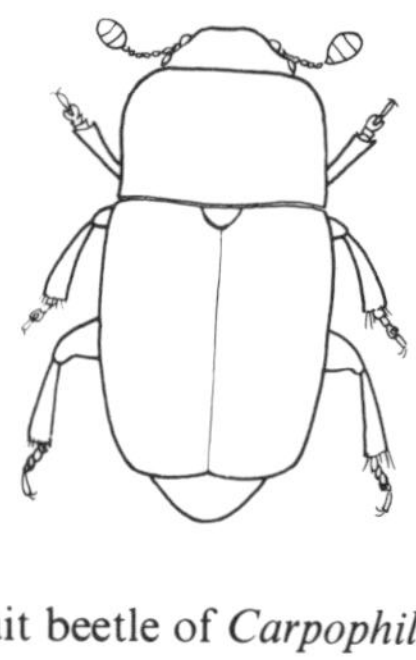

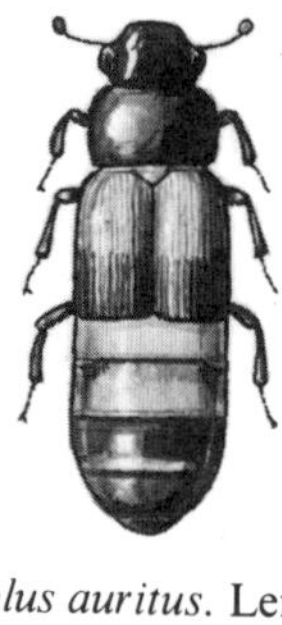

Brachypeplus auritus. Length 4 mm.

Fruit beetle of *Carpophilus* genus. Length 3 mm.

STEM BORERS (Family LANGURIIDAE)

Although only eight species have been described in Australia, about 40 stem borers are known throughout the world. They are small, rarely exceeding 10mm in length and have narrow cylindrically-shaped bodies well adapted for crawling inside the pithy stems of plants. The Australian species often have orange markings on the prothorax and elytra; the antennae are 11-segmented with three or four segments forming a club. The larvae are also long and cylindrical and are usually white.

Adults are found on flowers and may be collected by sweeping a net backwards and forwards while walking through high vegetation.

FUNGUS BEETLES (Families EROTYLIDAE, ENDOMYCHIDAE and MYCETOPHAGIDAE)

Found up to a length of 23mm, erotylid fungus beetles are hemispherical or elongated and may be seen sunning themselves on timber in fine weather. The antennae have 11 segments and a club of three flattened segments. Most adults have striking red, yellow and black markings. Adults live beneath the bark of trees during winter. The larvae, which are white to pale-brown cylindrical grubs, feed on fleshy fungi or on fungus growing in decaying wood.

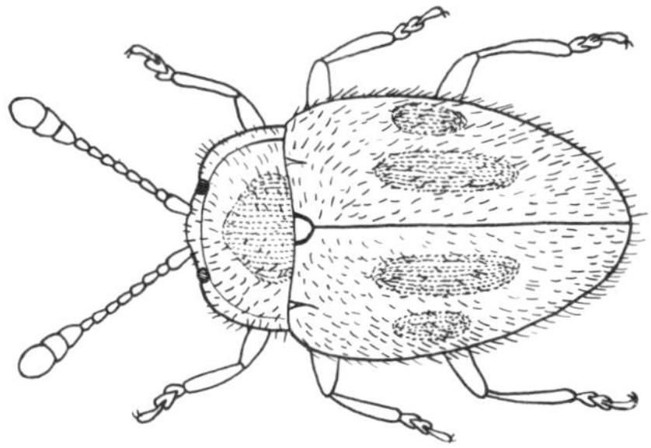

Fungus beetle of Endomychidae family. Length 3 mm.

Australia's largest species is the 25mm long species, *Episcaphula hercules*, which is found in the rain forests of North Queensland.

Sometimes known as 'handsome fungus beetles', the Endomychidae are smaller and more circular in shape than the preceding family, giving them something of the appearance of ladybirds. The 32 Australian species range from 2 to 8mm in length and are usually found in timbered country, around old logs and dead trees where they eat fungi. They may be identified by globular coxae on the first and second pairs of legs and by the four-segmented tarsi on all feet.

Beetles in the family Mycetophagidae are also known as fungus beetles and are very small, rarely exceeding 3mm in length, and are oval and often covered with a velvet down. Adults are often found feeding on fungi.

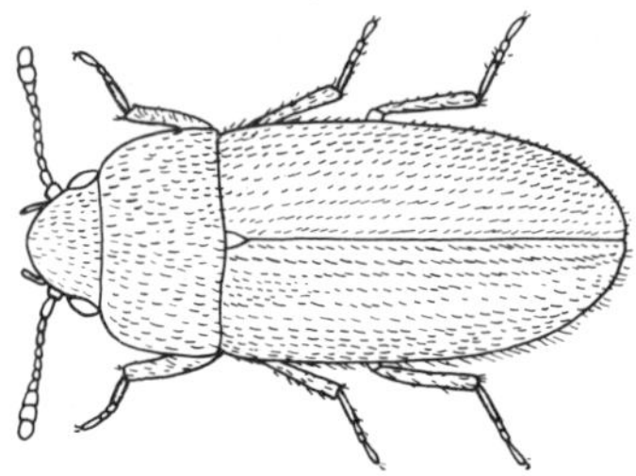

Fungus beetle of Mycetophagidae family. Length 3 mm.

LADYBIRDS (Family COCCINELLIDAE)

With their brilliant colours and contrasting spots, ladybirds are one of the best-known beetle families. Their colours may range from black, blue and green to the more familiar yellow and orange. To date 260 species have been described in Australia and all have three-segmented tarsi, and toothed claws.

The larvae have well formed legs and are often spiny. Like the adults, larvae are mostly predators of aphids and other small insects and mites. Some species in the subfamily Epilachninae eat leaves and may cause damage to vegetables.

However, the majority are beneficial to man, perhaps the most famous instance being the successful introduction of the Australian cardinal ladybird *(Rodolia cardinalis)* to California where it saved the citrus industry from destruction by the cottony cushion scale *(Icerya purchasi)*. In New Zealand, the Australian black ladybird *(Rhizobius ventralis)* was used with equal success in combating the gum-tree scale *(Eriococcus coriaceus)* which was damaging blue gum plantations. Another Australian species, the mealybug ladybird *(Cryptolaemus montrouzieri)* was used to combat pests in Hawaii and California. Its larvae look remarkably like the mealybugs on which they prey.

Ladybirds can emit a yellow and pungent secretion from the joints of their legs and this is believed to be a protection against insect eaters.

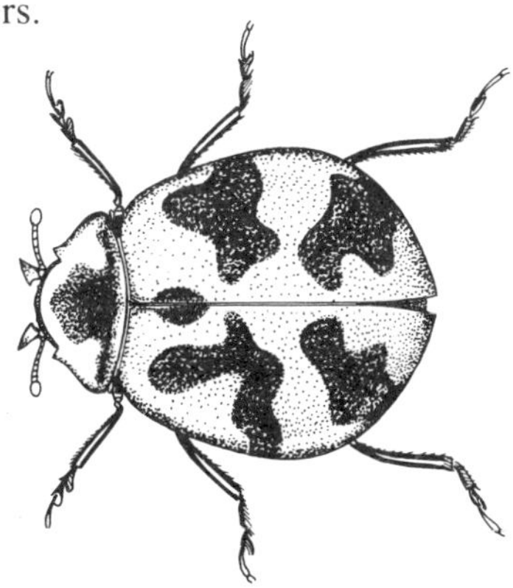

Ladybird.

TENEBRIONID BEETLES (Family TENEBRIONIDAE)

With 1220 species described in Australia and more than 15 000 species throughout the world, this is one of the largest beetle families and includes insects of many varied shapes. The tarsal segments however, are always five-five-four and adults usually have a thick outer covering. Larvae are usually shaped like half-cylinders with short legs, the forelegs being slightly longer than the others.

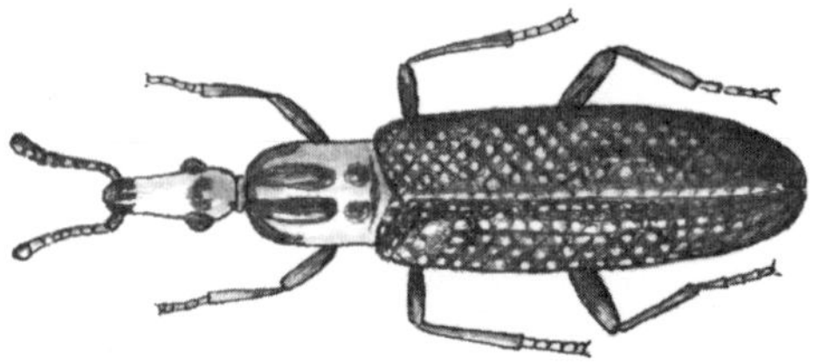

Antnest dwelling tenebrionid of the *Tretothorax* genus. Length 11 mm.

Many species are found beneath stones, logs and bark. The pest species, the yellow mealworm *(Tenebrio molitor)* is a nuisance in stored grain products while its larvae, mealworms, are bred and sold as food for reptiles and birds.

The heavily sculptured, 25mm long ironbark beetle *(Zopherosis georgei)* is dark brown. Its outside is so hard that an ordinary entomologist's pin is unlikely to penetrate it.

The ironbark beetle.

Species in the genera *Sphargeris*, *Caediomorpha* and *Caedius* are found on beaches.

Species in the subfamily Cyphaleinae are often up to 20mm long and may be brilliantly coloured, while those in the Adeliinae will emit a defensive odour when disturbed.

Undoubtedly the most interesting beetles in this family are the 'pie-dish beetles' in the genera *Helaeus*, *Pterohelaeus* and *Sympetes*. They are remarkably flattened with a broad flange around the outer margin of the elytra which are fused together. Pie-dish beetles have no flying wings and some of the most interesting species have been found in dusty places in the arid interior of Western Australia, and in Queensland.

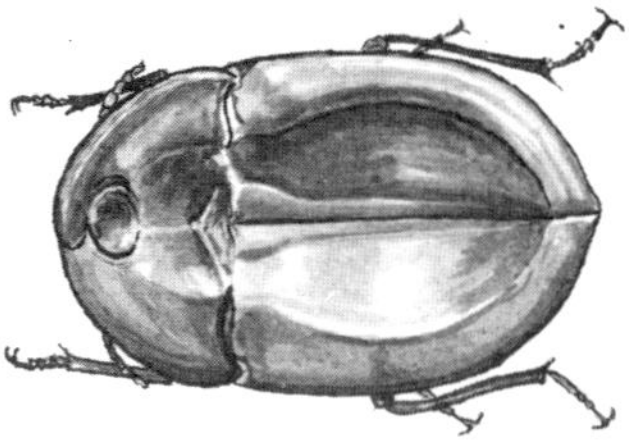

A pie-dish beetle, *Helaeus brevicostatus*. Length 35 mm.

PIN-TAIL BEETLES (Family MORDELLIDAE)

Growing up to 13mm in length, these humped beetles have the abdomen extending beyond the elytra and tapering to a long, stout spine. Adults frequent flowers and may jump when disturbed.

BEE PARASITES (Family RHIPIPHORIDAE)

Somewhat wedge-shaped and growing to about 30mm, these beetles are brown or black with yellow, brown or black patterns on their elytra. Males have well developed wings but reduced elytra and fanlike antennae. The females of these species are often wingless and have no elytra.

While adults are commonly found on flowers, the larvae of the subfamily Rhipiphorinae are parasites of wasp larvae (Scoliidae and Tiphiidae); those of the Rhipiphorini parasitise the larvae of bees; and the Rhipidiinae are parasites of cockroaches.

The eggs are laid on flowers, and the hard-covered oval larva must remain until a bee or wasp alights. It then attaches itself to the host and is carried to the nest or host larva. Then it penetrates it and feeds from within. After the first moult, the larger beetle larva emerges and moults into a legless grub which curls around the thorax of the host larva and feeds on it from the outside.

BLISTER or **OIL BEETLES** (Family MELOIDAE)

With rather soft covering, a reddish-yellow to black head and pronotum and a reddish-yellow to blue, green or black elytra, these beetles have a distinct head, with a neck separating it from the pronotum which is narrower than the elytra. The adults are plant eaters but larvae are parasites or prey on eggs and stores of bees, and the egg pods of Orthoptera. The adult blister beetles contain the chemical cantharidin which is poisonous and also causes blisters. Deaths have occurred from eating meloids and drinking liquids in which they may have been accidentally immersed.

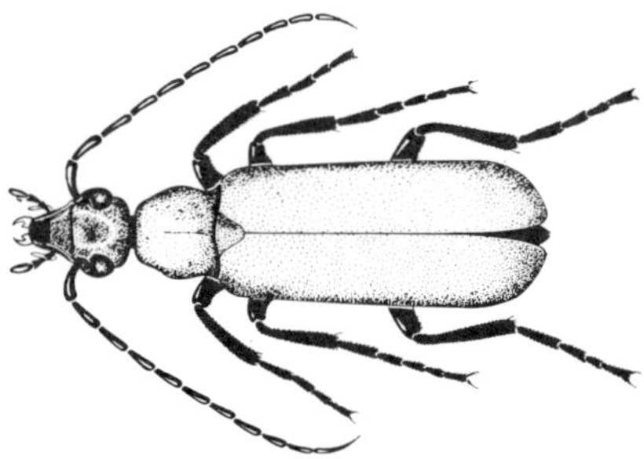

Blister beetle.

LONGICORN BEETLES (Family CERAMBYCIDAE)

Conspicuous for their long antennae and usually long slender bodies, the longicorn beetles are a major Australian group with 1000 of the world's 20 000 described species. The largest, *Titanus giganteus* from Brazil and the Guayanas, attains a length of 150mm. The antennae, which in males often exceed the length of the body, are set on prominences which are close to the eyes and possess a ball-joint. All larvae are wood borers and are yellowish grubs with small heads and large thoracic segments; their legs are usually small. The females either bore holes for their eggs in dead or decaying trees or deposit them in existing crevices.

Longicorns are divided into three groups. The Prioninae are usually large insects, measuring up to 50mm in length.

Often these longicorns have wider, shorter bodies and may have teeth on the pronotum. One of the most common is the chestnut brown species *Eurynassa odewahni*, which attacks eucalypts, while the species *Paroplites australis* infests banksias. Broader bodies are found among the genus *Sceleacantha*.

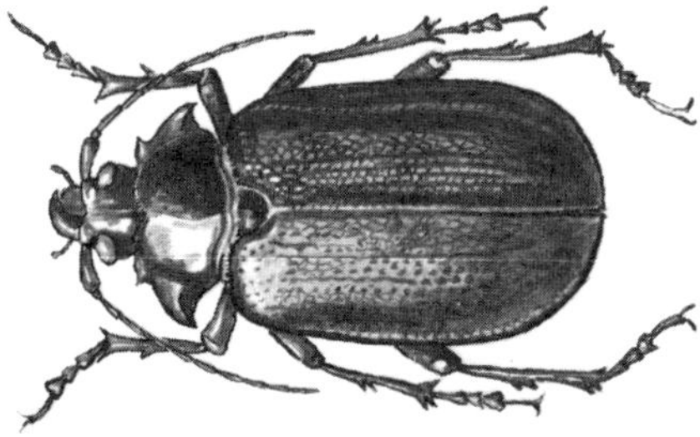

An example of the broad-bodied *Sceleacantha* genus, female. Length 45 mm.

The dull green *Rhipidocerus australasiae* rarely attains a body length of 20mm and is found between New South Wales and North Queensland. The species *Phaolus metallicum* is a vivid blue.

The second group, the Cerambycinae, are usually slender insects with no margins on the pronotum and no grooves in the fore tibia. One of the most conspicuous species is the yellow and black common eucalypt longicorn *(Phoracantha semipunctata)* which also has spined antennae and a spine on each side of the prothorax at the base of each elytron. Acacias are attacked by *Uracanthus triangularis*, a slim beetle, basically grey or light brown with a glossy triangle of brown on each wing cover.

The species in the genera *Sternaderus*, *Eroschema* and *Pterostenus* have a very strong similarity to beetles in the family Lycidae and species of *Macrones* have bodies with slender waists like sphecid wasps. Males in the genus *Piesarthrus* have huge decorative antennae.

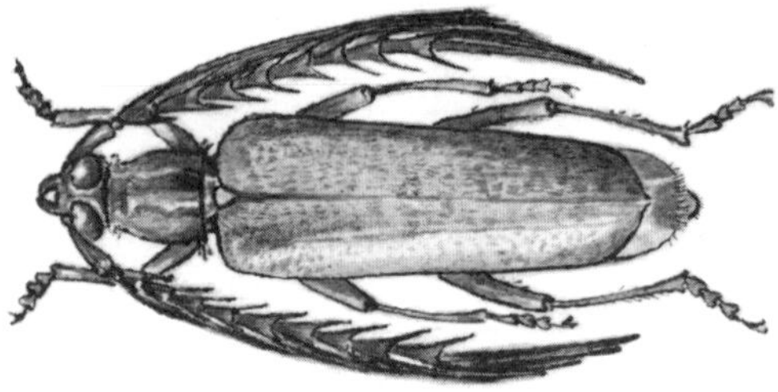

Male of the *Piesarthrus* genus. Length 40 mm.

The third group of longicorns are the Lamiinae. They include the North Queensland species *Batocera boisduvali* which infests the stems of native figs. It has a body up to 60mm long and antennae up to 130mm long. Its elytra are dark-grey to black but often spotted with yellow or white.

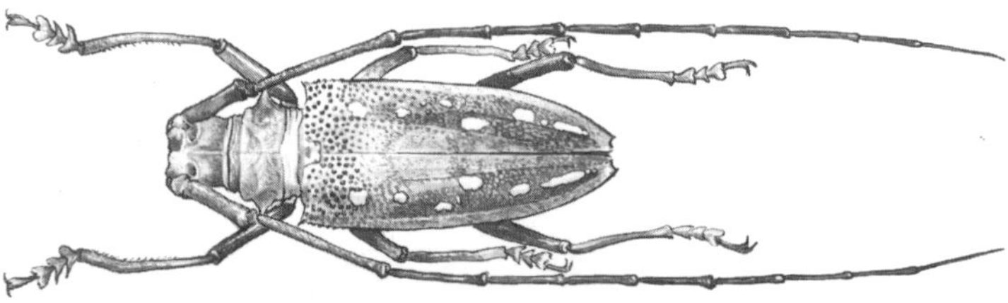

Batocera boisduvali. Body length 60 mm.

An even larger species *B. wallacei* found in New Guinea and occasionally in northern Queensland, has a body length of 75mm and antennae which, when pulled into a straight line, are said to have a span of 430mm. Also in this group are certain ground and bark-dwelling species which are often brown and resemble the body form of weevils but with long antennae.

PEA and **BEAN PESTS** (Family BRUCHIDAE)

With a head projecting as a distinct muzzle, these small dull brown beetles (1·5 to 6mm long) lay their eggs in the pods of peas and beans. The adult of the pestiferous pea weevil, *Bruchus pisorum*, feeds on the plants, while its larvae live in the seeds. While this species lives only on live peas, the species *Acanthoscelides obtectus* will attack both growing and dried peas and beans.

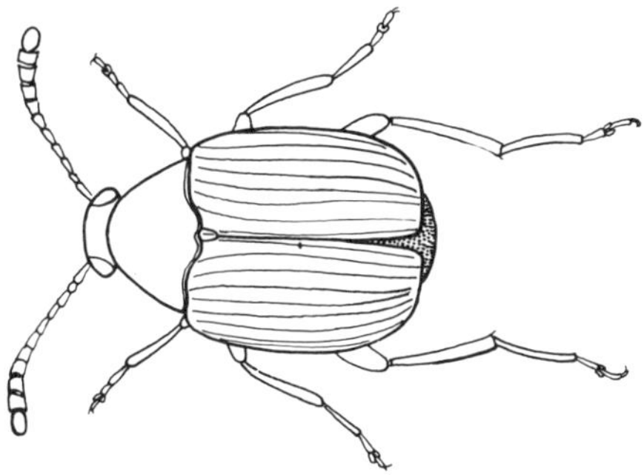

Example of *Bruchus* genus. Length 3 mm.

LEAF BEETLES (Family CHRYSOMELIDAE)

To date, 2100 species of leaf beetles have been described in Australia (there are some 20 000 in the world). These beetles may grow to a length of 22mm but many are much smaller. In such a large family, there are many different forms but all are more or less oval or rounded with short to medium-length antennae. The Australian species are grouped into 10 of the world's 16 subfamilies which include the potato pest Colorado beetle which has been excluded from Australia.

Leaf beetles are often brightly coloured with red, green and gold although these colours tend to fade to yellow brown after death. Larvae and adults are most commonly found on gum trees.

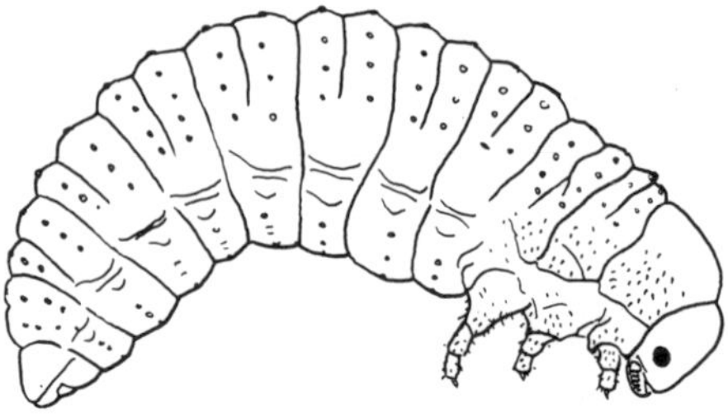

Paropsis larva. Length 15 mm.

Larvae of the eucalyptus tortoise beetle *(Paropsis dilatata)* eat eucalypts and wattles. Adult beetles of this species have paired glands which produce hydrocyanic acid and which are used for self-defence. The wattle-feeding larvae of *Paropsis orphana* are olive green and cling to wattle stems.

Grass beetles in the genus *Rhyparida* (subfamily Eumolpinae) feed on grasses and attack sugar cane. The black leaf beetle, *Rhyparida morosa*, is hemispherical and shiny black in colour. It has suckers on its legs which enable it to climb vertically. When disturbed however, it drops to the ground. The larvae tunnel in the stems of the host plants and pupate in the soil. Another beetle in this subfamily is the species *Spilopyra sumptuosa*, a brilliant green and purple insect found in northern New South Wales.

Beetles in the subfamily Galerucinae have the bases of the antennae set very close together. A number of pests are included in this group including the red-shouldered leaf beetle *(Monolepta australis)*, a red and yellow species which attacks maize, stone fruit and citrus trees. The pumpkin beetle, *Aulacophora hilaris* and related species destroy the foliage and flowers of cucumbers, melons and pumpkins.

The bright green *Hoplostenes viridipennis* can feed without injury on the leaves of the stinging tree *(Laportea gigas)*. The reddish-brown fig-leaf beetle *(Galerucella semipullata)* which grows to 10mm inhabits and lives off the Moreton Bay and cultivated fig trees.

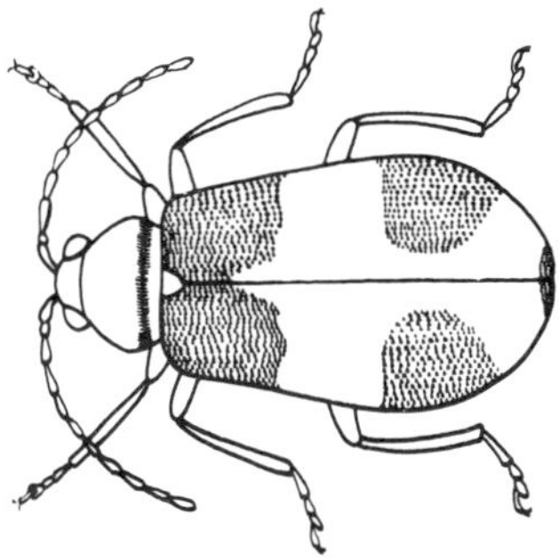

Aulacophora hilaris, the pumpkin beetle. Length 7 mm.

Flea beetles in the subfamily Halticinae, as their name suggests, have the femora of the hind-legs adapted for jumping in much the same way as fleas. Some can move over half a metre in a single bound. They are small and metallic and attack many plants, causing characteristic small holes in the foliage.

The North Queensland metallic green kangaroo beetles *(Sagra papuana)* grow to 20mm and also have large hind-legs, however they appear unable to jump.

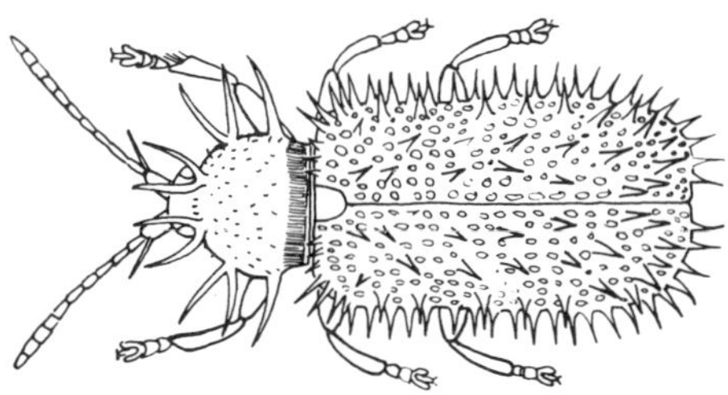

Monochirus multispinosus. Length 5 mm.

Adapted to a life within the growing stems of green plants, beetles in the Hispinae subfamily are generally elongated with short stiff protruding antennae. Their larvae also feed within the stems. The species *Monochirus multispinosus* is a black beetle found in grassy swamps in New South Wales.

The tortoise beetles (subfamily Cassidinae) have the head concealed beneath the prothorax and appear almost circular from above. The elytra have a strong flange which fits close to the leaf on which they rest. They may be yellow and black with spots or patches of iridescent gold, however this colour is produced from body fluids and disappears when the beetle dies. Most Australian species are included in the genus *Aspidomorpha*. The larvae are leaf eaters and carry their cast skins on their backs using long spines.

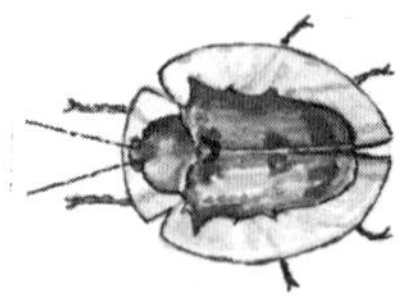

Tortoise beetle of the *Chirida* genus. Length 13 mm.

Weevils

Superfamily CURCULIONOIDEA

With 4300 species described in Australia, the weevils are a very large group although all are linked by the possession of an elongated beak or snout *(rostrum)* which has earned some the name 'elephant beetles'. The beak may be short or long, broad or slender. It may curve down or point ahead. On all weevils, however, the mouth and jaws are at the end of the beak. The body is usually roughened with various forms of knobs, bristles or scales. The larvae do not usually possess legs; and, like the adults, they can be serious pests of cultivated crops and stored foods.

Fungus weevils in the family Anthribidae have snouts of varied shapes and lengths and antennae which may or may not be clubbed. Legs have four tarsi which appear to be three. These weevils are generally small and brown but some species attain a length of 15mm. The species *Doticus pestilans* breeds in the woody galls caused by rust fungus in wattles. This weevil will also inhabit apples left on trees until they dehydrate. The species *Araecerus fasciculatus* attacks coffee, cocoa and spices and is found in all warmer parts of the world.

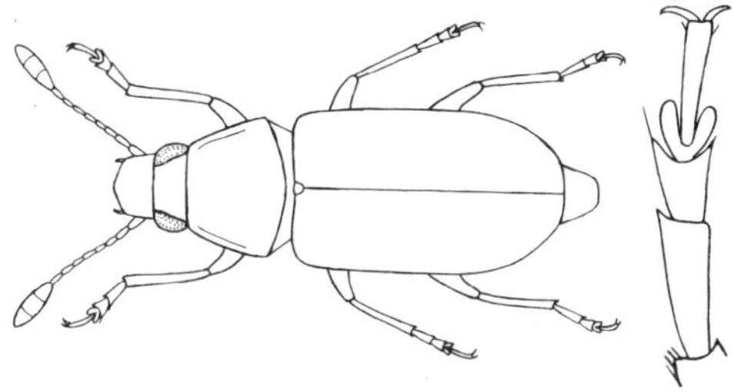

Anthribid weevil, length 6 mm. Also, enlarged hind tarsus.

Primitive weevils in the family Brenthidae may grow to 40mm in length and are long and parallel-sided with an elongated and slender head and snout. The short antennae are located some distance up the snout. Larvae are unusual as they have small legs. Both adults and larvae are thought to feed in decaying wood. They are often reddish and species in the genera *Ectocemus* and *Cyphagogus* are sometimes found under eucalyptus tree bark, more often in the warmer parts of Australia. The species *Cordus hospes* lives in the nests of ants and termites.

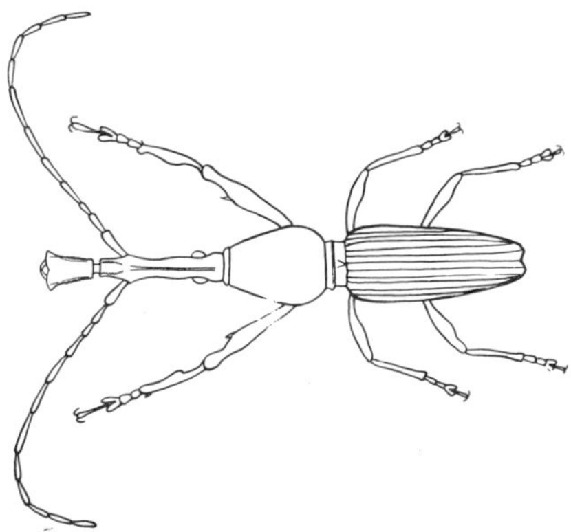

An example from *Ectocemus* genus. Length 25 mm.

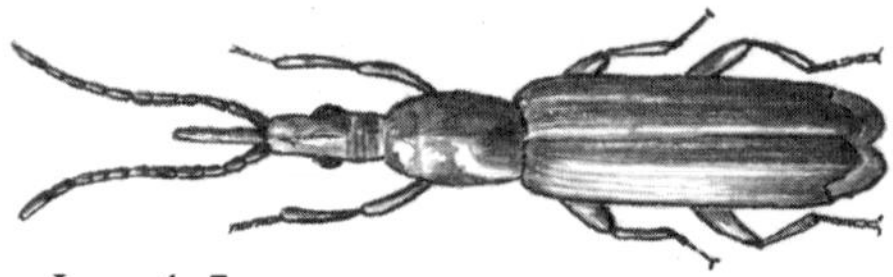

Cordus hospes. Length 7 mm.

The species *Apion convexipenne* in the family Apionidae is 4mm long and is a strange-looking weevil with a short humped body, a long curved snout and long legs. The introduced sweet-potato weevil *(Cylas formicarius)* is red with black markings.

In the major family of weevils, the Curculionidae, there are insects varying in adult length up to 50mm; there are 4000 species described in Australia. Most species have the head extended into a snout and the mouth parts are always at its tip. Antennae are elbowed and usually have a compact three-or four-segmented club. Larvae are small curl-grubs, legless and usually feeding in hidden locations. They pupate within the food plant or underground.

Ground dwelling weevils are large, spiny and usually found beneath stones and logs. They are wingless and in the genus *Phalidura*, males have large projecting forceps at the end of their abdomen.

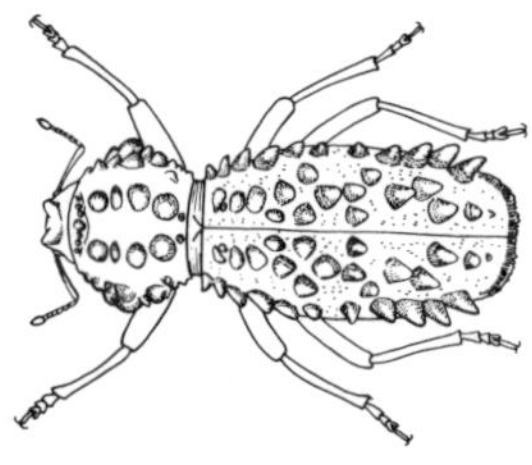

Ground dwelling weevil of *Macramycterus* genus.

The diamond weevils *(Chrysolophus spectabilis)* are a brilliant metallic blue or green. They were first collected by Sir Joseph Banks at Botany Bay and their larvae feed on the wood of various acacia trees.

The giant pine weevil, *Eurhamphus fasciculatus*. Length 65 mm.

As its name suggests, the larva of Australia's largest weevil, the giant pine weevil *(Eurhamphus fasciculatus)* tunnels in the hoop pine. Adults attain a length of up to 50mm and it has a very long snout with distinct dark brown clumps of bristles on the pronotum and elytra. Also in the subfamily Curculioninae, the elephant weevil *(Orthorrhinus cylindrirostris)* grows to 20mm long and has a long straight downward-pointing snout with antennae near the tip. It has disproportionately long front legs with very broad pads on them. It is now a pest of fruit trees and its larvae tunnel into branches. A smaller relative *O. klugi* damages grape vines.

Wingless weevils in the subfamily Leptopiinae include the apple root weevil *(Leptopius squalidus)* which lays eggs on leaves and its grubs, after hatching, crawl down to the roots where they feed, gouging deep furrows which may cause the roots to die. Adults feed on young leaves and shoots. The stout oval or spiny weevils in the genus *Catasarcus* are found mainly in Western Australia.

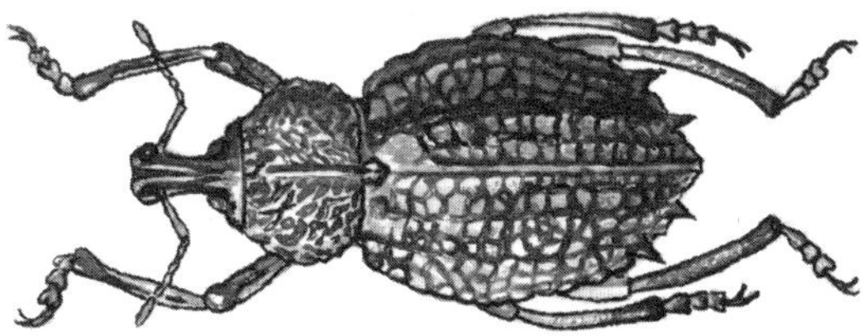

Example of the Leptopiinae.

Equally notorious are the dicky rice weevil *(Maleuterpes spinipes)* a small (3mm long) brown-grey pest of citrus, and the introduced Argentinian white-fringed weevil *(Graphognathus leucoloma)* which may affect many cultivated plants.

In Queensland the subfamily Otiorrhynchinae includes the metallic blue or green species in the genus *Myllocerus*. They are pests of fruits, trees and vines.

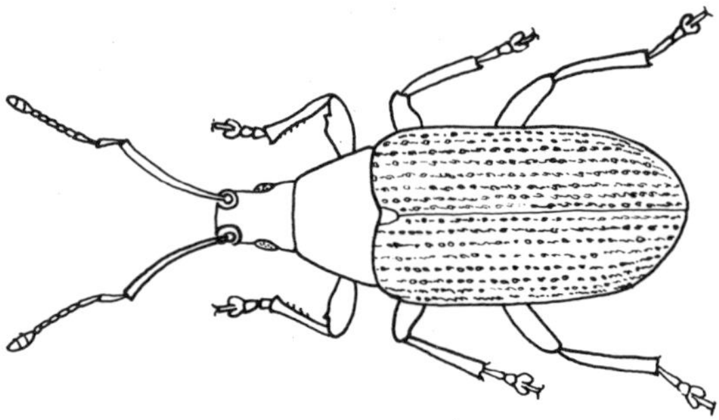

Example of the *Myllocerus* genus. Length 6 mm.

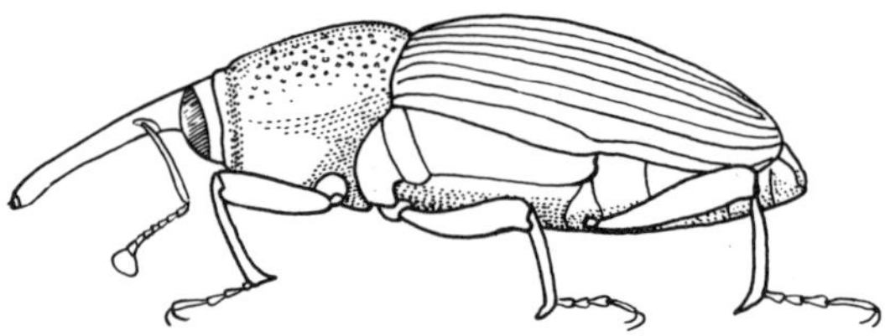

Trignotarsus rugosus.

Two large species of weevils are *Trignotarsus rugosus*, a heavily-marked dark species growing to 25mm with a longish snout, found in New South Wales and the 35mm long *Iphthimorhinus australasiae* found in northern Queensland. Also in this group are the grain pests in the genus *Sitophilus* which include the granary, rice and maize weevils and the banana weevil borer *(Cosmopolites sordidus)* of Queensland.

The little bark weevils are well-known for the designs they make on the bark and trunks of trees. They have cylindrical bodies, and a large antennal club. They burrow under the bark and the females lay their eggs in the tunnels. Young are raised in separate cradles or pits. To feed them, the adults cultivate fungus (ambrosia) which is grown among wood chips. When the larvae hatch and grow, they enlarge their individual pits and their excreta is removed by the female, some being used to fertilise the fungus bed. The opening to each pit is closed by her with a plug of fungus and when eaten, it is replaced with fresh food. In tropical rain forests, ambrosia beetles are more numerous than bark beetles. The opposite is the case in temperate forests.

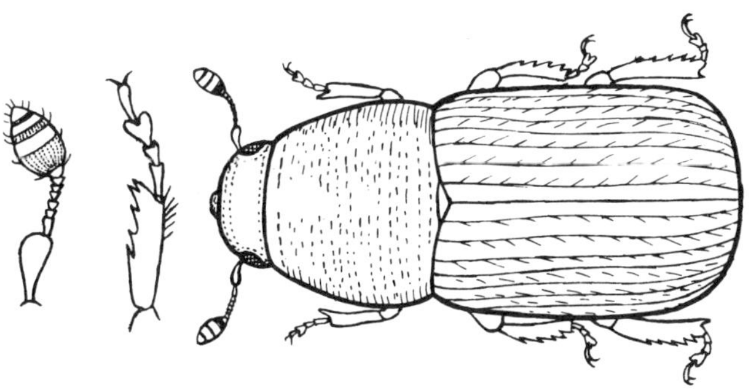

Ambrosia beetle. Also, antenna and hind leg enlarged.

STYLOPS

Order STREPSIPTERA (twisting wings)

Although only nine species have been described from Australia, 93 species of stylops are now known here of the 300 so far recorded throughout the world. These insects are mostly parasitic on other insects and spend most of their life-cycle within the host. They are small insects with adult males attaining a maximum length of from 1·5 to 4mm with large wings while the females remain wingless and live within the pupal cases inside the host.

Stylops are the only insects which do not possess a second leg joint *(trochanter)* on adults. Most females also develop a unique brood canal—a gap between the body and the retained pupal case. Females resemble the larvae of certain beetles, although they can be found only if the host is carefully examined. The flattened disc-like head may be seen on the insect's abdomen.

Males have forewings which are reduced like the haltares of Diptera (see p. 135) but the hind-wings are greatly expanded and extend to three quarters of the abdominal length when fully spread. There are few veins in the wings in relation to their overall size.

Males of most species emerge in free flight on sunny days in late summer and despite their small size, may be noticed because of their pale, translucent wings. They fly in a strange, upright position, although the upturned end of the abdomen makes them seem to be at a horizontal angle of flight.

Males and females develop as parasites, and females live in the abdomen of certain species of leaf-hoppers (Hemiptera) and in some bees and wasps. The number of stylops seems to vary between hosts but up to five females have been collected from a single insect. After emerging from the host, males actively seek the females which remain only partly visible on their hosts. Some males have expanded forelegs to help grasp the female's cephalothorax when mating.

Larvae hatch inside the female stylops' body and several

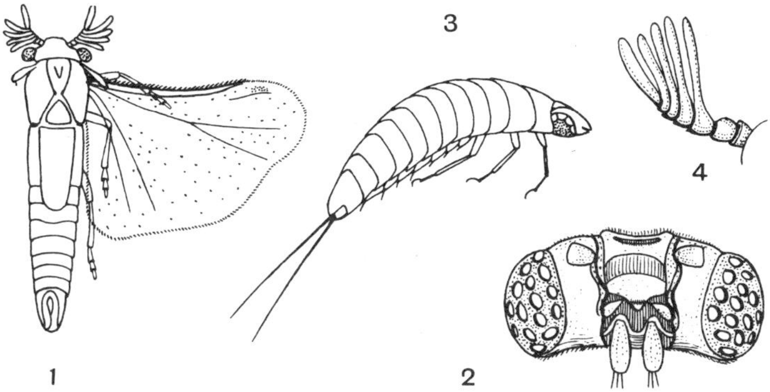

1. Stylops of Halictophagidae. 2. Stylops head. 3. Stylops larva. 4. Antenna.

thousand emerge through the brood canal and remain on the host's body until they can escape. These minute, active larvae have well-formed eyes and legs and long cerci at the tail. It is assumed they must be carried to the host's nest so that they can reach its larvae; they possess special enzymes which allow them to penetrate the larval skin of the host. Once inside the larvae, the stylops lose their legs and become like maggots, feeding by filtering the host's blood through their own skin. The stylops never penetrates the host's organs but occupies space between them, pushing them out of position. When mature, the stylops' head emerges and protrudes from the host's abdomen, often at the joint between segments. Hosts do not readily show they contain stylops although the wings on some bugs may be slightly displaced when they are present.

SCORPION FLIES

Order MECOPTERA (long wings)

These predatory, long-winged insects are more easily described because of their superficial similarity to other orders.

Some may look and fly like lacewings (Neuroptera); others are similar to crane flies (Diptera) but have four wings; and the genus *Austromerope* resembles a small cockroach. Scorpion flies have long spindly legs and long transparent veined wings, usually with the front pair longer than the hind-wings. At rest, the wings are held like a roof over the abdomen. The head is produced into a short beak and the antennae are long and thread-like *(filiform)* with many segments. The compound eyes are well developed.

The term 'scorpion fly' refers to the males of some overseas species which fly with the tip of the abdomen raised and the two tail claws resembling the stinging spine of a scorpion. However this has not been reported from any of the 20 species known in Australia. Larvae of most Australian species are terrestrial and resemble caterpillars but those of the genus *Nannochorista* are aquatic and resemble elongated wire-worms. Larvae possess compound eyes.

Scorpion flies prefer cool damp locations and *Nannochorista* adults are usually seen around marshes associated with lakes and mountain streams. Other species fly among blossoms and prey on other insects including flies, bees, caterpillars, fireflies and the larvae of other beetles. Some species hatch in spring but others only in autumn.

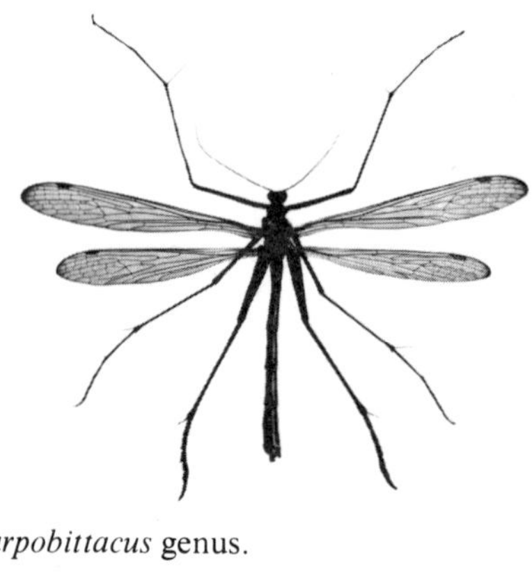

Example of *Harpobittacus* genus.

Courtship is an involved and prolonged process. In the common genus *Harpobittacus* it begins by the male securing prey with its hind-legs. It lands on a low shrub where the prey is carried to the mouth. Attractants released by the male bring a female who is grasped by the male's hind-legs. The female is released and is then fed during mating by the male who transfers the 'nuptial feast' from his mouth to hers with his legs. Mating usually takes place in hot weather.

Females will only mate with males possessing food and one male may attract several females in turn, sharing a single item of food with each. It has been suggested that animal food is needed to help eggs develop because females, when alone, obtain nectar but are not known to be predatory. Eggs may be laid in the ground, be dropped at will, or deposited among grass.

Larvae may take up to three months, or until rain falls, before emerging from eggs. Larvae are also predatory and pupate just below the surface of the soil.

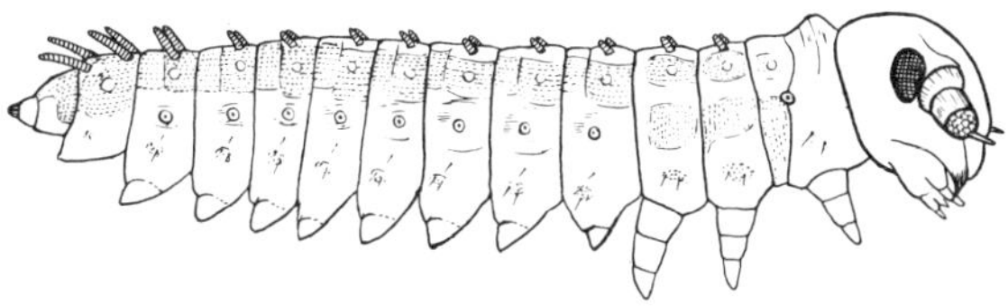

Larva of *Chorista australis*.

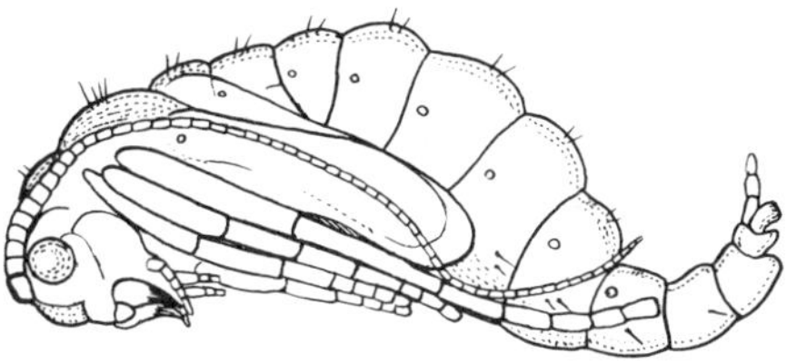

Pupa of *Chorista australis*.

Five families of scorpion flies are represented in Australia and identification is based mainly on wing venation. The species *Chorista australis* is found in most inland parts of south-eastern Australia and in coastal New South Wales. Its larvae are pink and live in tunnels in soft damp soil.

Species in the family Bittacidae are red and black and those in the genus *Harpobittacus* may attain a wing-span of 50mm Larvae of this genus may be nearly 25mm long and have caterpillar-like spines on the body.

FLEAS

Order SIPHONAPTERA (tube wings)

With their wingless, flattened bodies, hind-legs adapted for jumping *(saltatorial)* and piercing-sucking mouthparts, fleas are one of the most persistent pests of man and other warm-blooded vertebrates. They are also carriers of diseases including bubonic plague and murine typhus which they can transmit to man; fleas also act as intermediate hosts of tapeworms.

Fleas are highly adapted to living in their hosts' hair or feathers. Their flattened bodies enable them to slide easily between hairs or feathers while backward-pointing bristles or hairs, sometimes arranged in combs *(ctenidae)* assist forward progress and hamper attemps to dislodge them.

While various species of fleas are adapted to particular hosts, many show far greater adaptability than lice (see pp. 45–9) and will move to a different host if the preferred one is not available. Fleas can also withstand extremely unfavourable conditions for long periods. Human fleas *(Pulex irritans)* can live for 125 days without food and for more than 500 days on human blood.

Flea eggs look similar to fine white sugar and may be seen in pets' kennels or baskets which have not been dusted regularly with insecticide. Larvae are small white legless maggots up to 10mm long. Larvae are not bloodsuckers but feed on scales of skin and other plant or animal debris which collects where they hatch. They pupate in round, flat silken cocoons and pupae may remain for long periods in this stage. It has been suggested that vibration is the stimulus needed for adults to emerge and this is believed to explain the sudden appearance of fleas in houses which have been empty for long periods.

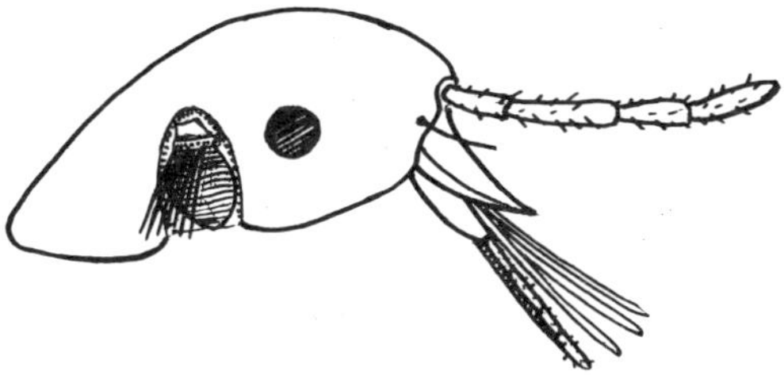

Head of *Pulex irritans*, the human flea.

Of the 1400 species of fleas described throughout the world, 100 are known in Australia including nine cosmopolitan species and three others known to be pests of penguins and petrels. Native fleas are often closely associated with marsupials and rodents, although some have related forms on similar fauna in South America. Some Australian bat fleas are also recorded.

Adults of both sexes are bloodsuckers and if a host dies, they leave it promptly and seek a new one. While one Australian species *Bradiopsylla echidnae* has been found only on echidnas, at least one other species of flea has been recorded from as many as 24 different types of host animal. Also, while some animals will attract only one species of flea, the southern bush rat *(Rattus fuscipes)* has had 11 different species of fleas identified on it.

Other fleas may inhabit birds' nests and those of some birds may have a life-cycle closely linked with the host's.

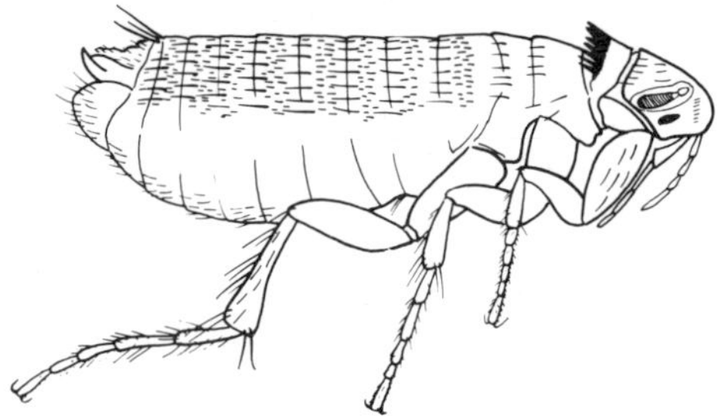

Pygiopsylla hoplia, common flea of Australian marsupials especially native cats. Length 2 mm.

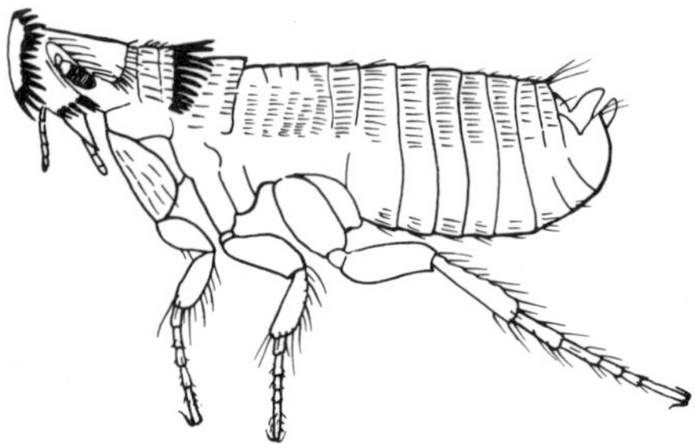

Stephanocircus dasyuri, common flea of Australian marsupials, especially bandicoots. Length 1.8 mm.

Australian fleas are grouped into nine families and identification is based on various anatomical features including the arrangement of bristles on the body. The main introduced species are the human flea, dog fleas *(Ctenocephalides canis)*, cat fleas *(C. felis)* and rat fleas *(Xenopsylla cheopis)*. All can be found on a variety of hosts. Stickfast fleas (genus *Echidnophaga)* differ from other fleas by attaching themselves to the host and sucking blood continuously. The poultry stickfast flea *(E. gallinacea)* is a serious pest of fowls in many parts of the world and heavy infestation can cause loss of condition and reduced egg output.

PLATE 1

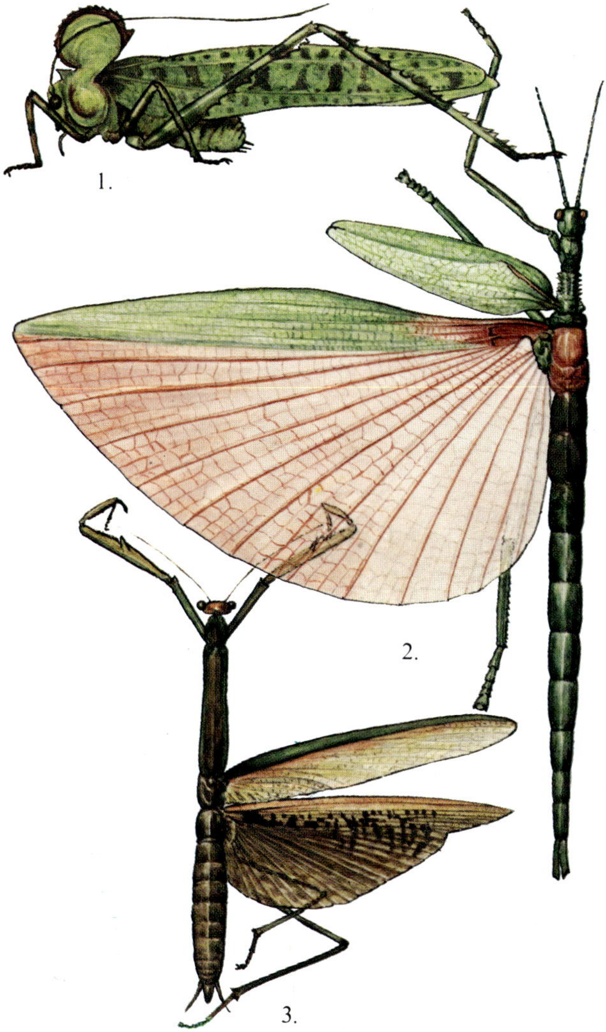

1. *Alectoria superba* Brunn. (Fam. TETTIGONIIDAE). 2. *Podacanthus typhon* McCoy (Fam. PHASMATIDAE). 3. *Tenodera australasiae* Leach (Fam. MANTIDAE).

PLATE 2

1. *Orthodera ministralis* Fabr. (Fam. MANTIDAE). 2. *Polyzosteria mitchelli* Angas (Fam. BLATTIDAE). 3. *Caedicia olivacea* Brunn. (Fam. TETTIGONIIDAE). 4. *Mastotermes darwiniensis* Frogg. (Fam. MASTOTERMITIDAE). 5. *Siphanta acuta* Walk. (Fam. FLATIDAE). 6. *Sephena cinerea* Kirk. (idem). 7. *Callipappus australis* Mask. (Fam. COCCIDAE). 8. *Biprorulus bibax* Bred. (Fam. PENTATOMIDAE). 9. *Commius elegans* Don. (Fam. PENTATOMIDAE). 10. *Ptilocnemus femoratus* Horv. (Fam. REDUVIIDAE).

PLATE 3

1. *Diphlebia euphaeoides* Till. (Fam. AMPHIPTERYGIDAE), male. 2. *Austropetalia patricia* Till. (Fam. AESHNIDAE), female. 3. *Rhyothemis graphiptera* Sel. (Fam. LIBELLULIDAE), male. 4. *Rhyothemis resplendens* Sel. (Fam. LIBELLULIDAE), male. 5. *Eusthenia lacustris* Till. (Fam. EUSTHENIIDAE). 6. *Sigmatoneura formosa* Bks. (Fam. PSOCIDAE). 7. *Psychopsis elegans* Guer. (Fam. PSYCHOPSIDAE). 8. *Porismus strigatus* Burm. (Fam. OSMYLIDAE). 9. *Periclystus aureolatus* Till. (Fam. MYRMELEONTIDAE). 10. *Stenopsychodes melanochrysa* Till. (Fam. PSYCHOMYIIDAE).

PLATE 4

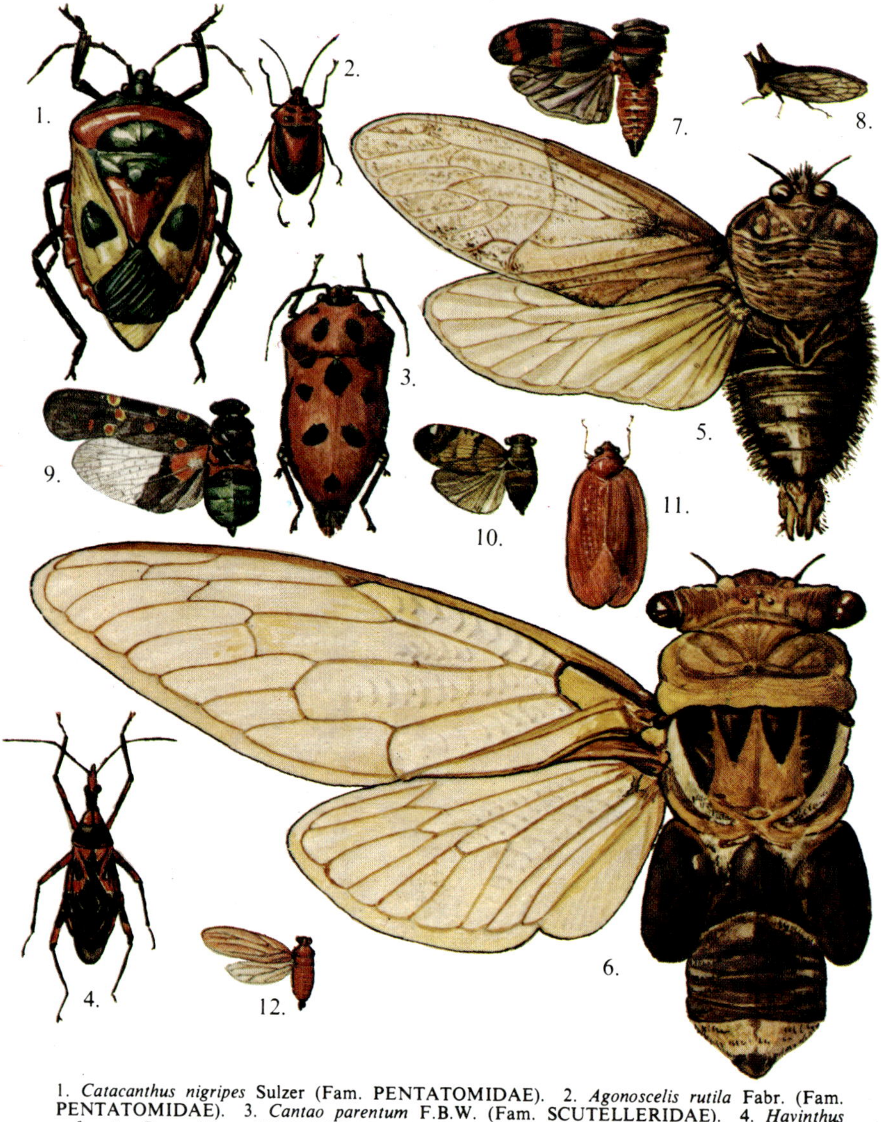

1. *Catacanthus nigripes* Sulzer (Fam. PENTATOMIDAE). 2. *Agonoscelis rutila* Fabr. (Fam. PENTATOMIDAE). 3. *Cantao parentum* F.B.W. (Fam. SCUTELLERIDAE). 4. *Havinthus rufovarius* Berg. (Fam. REDUVIIDAE). 5. *Thopha sessiliba* Dist. (Fam. CICADIDAE). 6. *Tettigarcta crinita* Dist. (Fam. TETTIGARCTIDAE). 7. *Eurymela rubrovittata* A. & S. (Fam. EURYMELIDAE). 8. *Eufairmairia fraterna* Dist. (Fam. MEMBRACIDAE). 9. *Desudaba maculata* Dist. (Fam. FULGORIDAE). 10. *Platybrachys decisa* Walk. (Fam. EURYBRACHYIDAE). 11. *Achilus flammeus* Kby. (Fam. ACHILIDAE). 12. *Creiis longipennis* Walk. (Fam. PSYLLIDAE).

PLATE 5

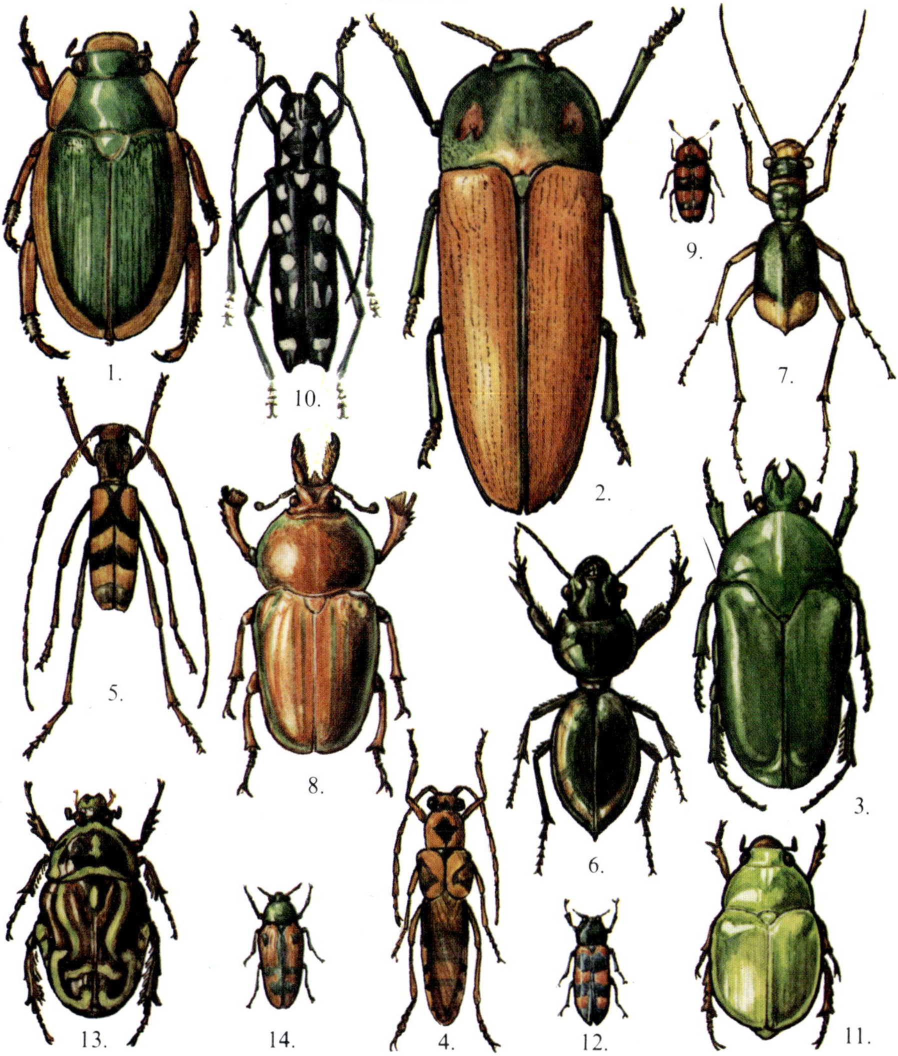

1. *Anoplognathus smaragdinus* Ohaus (Fam. SCARABAEIDAE). 2. *Calodema regalis* Cast & Gory (Fam. BUPRESTIDAE). 3. *Ischiopsopha yorkiana* Jans. (Fam. SCARABAEIDAE). 4. *Hesthesis ferruginea* Bd. (Fam. CERAMBYCIDAE). 5. *Aridaeus thoracicus* Don. (Fam. CERAMBYCIDAE). 6. *Conopterum spaldingi* Macl. (Fam. CARABIDAE). 7. *Megacephala australis* Chaud. (idem). 8. *Lamprima aurata* Latr. (Fam. LUCANIDAE), male. 9. *Episcaphula nigrofasciata* Blkb. (Fam. EROTYLIDAE). 10. *Glenea elegans* Fabr. (Fam. CERAMBYCIDAE). 11. *Anoplognathus mastersi* Macl. (Fam. SCARABAEIDAE). 12. *Melobasis gratiosissima* Thoms. (Fam. BUPRESTIDAE). 13. *Eupoecila australasiae* Don. (Fam. SCARABAEIDAE). 14. *Stigmodera scalaris* Bd. (Fam. BUPRESTIDAE).

PLATE 6

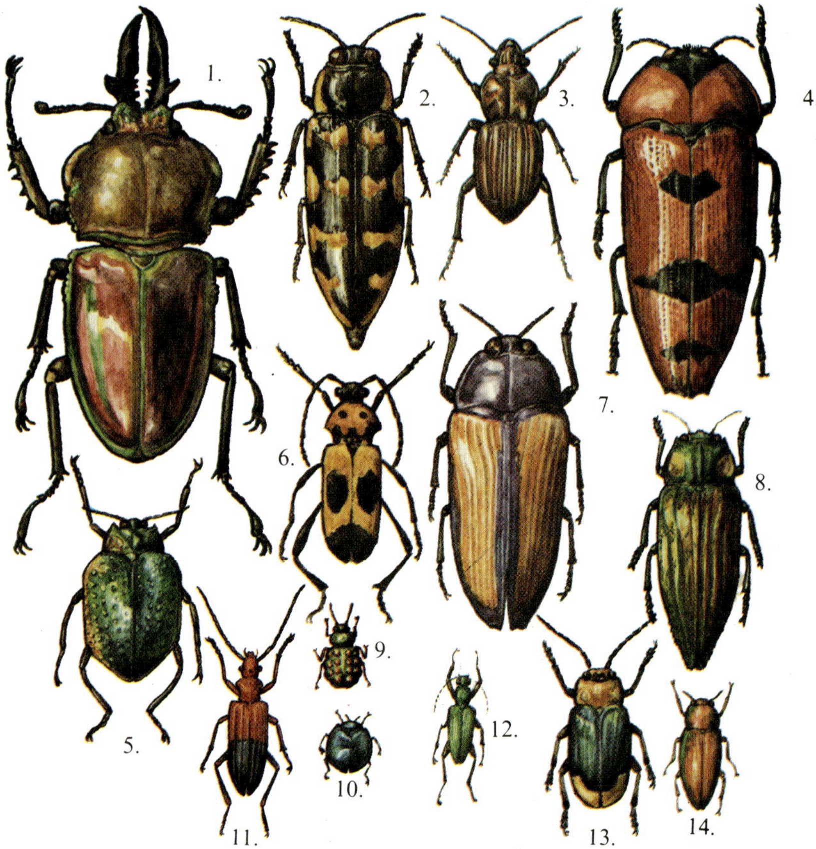

1. *Phalacrognathus muelleri* Macl. (Fam. LUCANIDAE), male. 2. *Cyria imperialis* Don. (Fam. BUPRESTIDAE). 3. *Notolestus sulcipennis* Macl. (Fam. CARABIDAE). 4. *Calodema plebeja* Jord. (Fam. BUPRESTIDAE). 5. *Prophanes mastersi* Pasc. (Fam. TENEBRIONIDAE). 6. *Purpuricenus quadrinotatus* Wh. (Fam. CERAMBYCIDAE). 7. *Stigmodera suturalis* Don. (Fam. BUPRESTIDAE). 8. *Chrysodema aurofoveata* Guer. (Fam. BUPRESTIDAE). 9. *Pedilophorus gemmatus* Lea. (Fam. BYRRHIDAE). 10. *Orcus chalybeus* Bd. (Fam. COCCINELLIDAE). 11. *Agasma semicrudum* Newm. (Fam. OEDEMERIDAE). 12. *Aethyssius viridis* Bd. (Fam. CISTELIDAE). 13. *Aesernia australica* Jac. (Fam. CHRYSOMELIDAE). 14. *Melobasis metallifera* Hope. (Fam. BUPRESTIDAE).

PLATE 7

1. *Oreixenica kershawi ella* Oll. (Fam. NYMPHALIDAE). 1. *Ornithoptera priamus euphorion* Gray (Fam. PAPILIONIDAE). 3. *Paralucia aurifera* Blanch (Fam. LYCAENIDAE), male. 4. *Euschemon rafflesia* W.S.Macleay (Fam. HESPERIIDAE). 5. *Graphiam macleayanum* Leach (Fam. PAPILIONIDAE).

PLATE 8

1. *Hypochrysops byzos hecalius* Misk. (Fam. LYCAENIDAE), male, underside. 2. *Graphium agamemnon ligatum* Roth. (Fam PAPILIONIDAE). 3. *Lucia limbaria* Fabr. (Fam. LYCAENIDAE), male. 4. *Doleschallia bisaltide australis* Feld. (Fam. NYMPHALIDAE). 5. *Vanessa itea* Fabr. (idem), N.Z. also. 6. *Cupha prosope prosope* Fabr. (idem). 7. *Deudorix epijarbas diovis* Hew. (Fam. LYCAENIDAE), male.

Goodwill
South Florida

Goodwill Industries #6104

BROWARD OUTLET
2104 W COMMERCIAL BLVD FT LAUDERDALE, FL 3
3309

TransId 6104-06104021783257698598
07/05/26 9:30 AM
Employee Kim N
Register: POS6104_2

HOUSE ACCOUNT

Outlet Books
1 @ $0.75 $0.75

Items:	1
Subtotal:	$0.75
Tax:	$0.05
Total:	$0.80
Cash:	$1.00
Change:	-$0.20

Thank you for shopping at
Goodwill South Florida!

Your purchase creates work opportunities
for people with disabilities.

Learn more: www.GoodwillSouthFlorida.org.

Find unique treasures at
www.ShopGoodwill.com.

All outlet sales are final.

6104-06104021783257698598

TRUE FLIES

Order DIPTERA (two-winged)

As their name suggests, these insects possess only one pair of transparent membraneous wings. The hind-wings are reduced to slender clubbed stalks known as balancers *(haltares)*. There are many insects included in this order and among them some of the best known are mosquitoes, crane flies and house flies.

All true flies have very large compound eyes which occupy most of the head and their mouthparts, while varying considerably, are always adapted for sucking, or piercing and sucking, but never for biting. However, unlike some other groups of insects which show wide diversity in appearance (are *heterogeneous*) Diptera are all very much alike *(homogeneous)*. There are exceptions and some Diptera are wingless *(apterous)*.

Metamorphosis is complete. Eggs are laid on plants, animals or in decaying matter or water and from these hatch legless larvae which are usually smooth-skinned maggots which taper sharply and have small heads. Among the more highly specialised forms, the last larval skin is not cast off but is retained as a protective case *(puparium)* under which the pupa is formed.

The maggot is specialised for a life in soft food masses and its narrow foreparts and broader, truncated hind parts enable it to wriggle through rotting, often semi-liquid material leaving only the broad tail exposed to gather air through two spiracles located there.

Pupae may be free as in aquatic forms (see pp. 234–7) or enclosed in a barrel-shaped case of hardened larval skin as with blowflies. In the blowfly pupa, the emerging insect has an inflatable bladder *(ptilinum)* on top of the head which is used to rupture the walls of the pupal casing, releasing the adult fly.

Most adults have bodies covered with hairs or bristles and antennae may assume many forms. Identification of species often rests on wing venation but broader groupings are shown in the following table.

It is estimated there are about 150 000 species in the world but Australia's recorded population is only 6800 species. These may be grouped in various ways.

Broad identification of Diptera may be made on the following features of adults. Only 29 of the 86 families present in Australia are included here.

Identification	*Common name*	*Family*	*See page*
Long antennae usually with 6–14 segments	Crane flies (daddy-long-legs)	TIPULIDAE	138–9
	Moth flies or midges	PSYCHODIDAE	139–40
	Mosquitoes	CULICIDAE	140–1
	True midges	CHIRONOMIDAE	142
	Biting midges	CERATOPOGONIDAE	142–3
	Black flies (true sand flies, buffalo midges)	SIMULIIDAE	142–4
	Net-veined midges	BLEPHARICERIDAE	144
	Gall midges or gnats	CECIDOMYIIDAE	144–5
	Fungus gnats	MYCETOPHILIDAE	145
Antennae with less than 6 segments	March, horse or gad flies	TABANIDAE	145–6
No ptiline fissure	Soldier flies	STRATIOMYIDAE	146
	Window flies	SCENOPINIDAE	146
	Robber or assassin flies	ASILIDAE	147–8
	Bee flies	BOMBYLIIDAE	148
	Water cruisers	EMPIDIDAE	149–50
	Stilt flies	DOLICHOPODIDAE	150
	Hover flies	SYRPHIDAE	151

Identification	*Common name*	*Family*	*See page*
Antennae with less than 6 segments	Wasp flies	CONOPIDAE	152
	Aust. stalk-eyed flies	PLATYSTOMATIDAE	152
	Fruit flies	TEPHRITIDAE	153–4
	Kelp flies	COELOPIDAE (Phycodromidae)	154
	Tomato flies	LONCHAEIDAE	154
	Leaf miners	AGROMYZIDAE	154
Ptiline fissure present above bases of antennae	Vinegar or ferment flies	DROSOPHILIDAE	155
	House flies	MUSCIDAE	155–6
	Blowflies	CALLIPHORIDAE	156–7
	Flesh flies	SARCOPHAGIDAE	157
	Tachinid flies		158
	Bot or warble flies	OESTRIDAE	158
	Louse flies or sheep keds	HIPPOBOSCIDAE	158–9

Many of the Diptera are dangerous to man, as carriers of malaria and yellow fever (mosquitoes), and of enteric diseases (house flies). Those which are economic pests include maggots in sheep (blowflies) and internal parasites in sheep, horses and kangaroos (bot flies). Fruit flies are serious pests of cultivated fruit and leaf miners and stem borers may also be of nuisance value to commercial crops.

However, many species are beneficial and pollinate plants, reduce weeds, and parasitise other insects which are themselves pests. Even mosquitoes have been useful in reducing the Australian rabbit population by transmitting myxomatosis and other species are being studied to determine whether they may be used to help mankind in a number of different ways.

CRANE FLIES (Family TIPULIDAE)

Common in swampy areas, in grasslands and along the banks of streams where they cling to vegetation, crane flies have long slender wings and bodies and exceptionally long spindly legs. They look like grotesquely large mosquitoes but are actually plant feeders. In flight, they are slow and clumsy with the front legs doubled while the others drag behind. The alternate term 'daddy-long-legs' is incorrect and should strictly be restricted to species of spiders frequently found inside homes which are grouped in the family Pholcidae, and live in tangled webs, often high in the corners of a room. Similarly, the terrestrial larvae of crane flies have been called 'wireworms' although this term should only be applied to the larvae of click beetles (Elateridae, see pp. 102–3). Crane fly terrestrial larvae are long and slender, and live in the ground feeding off roots of grasses and other plants. Aquatic larvae are described on p. 234. Pupae are long and remain in the soil until adult and ready to emerge. They then push themselves to the surface by using rings of spines on the abdominal segments.

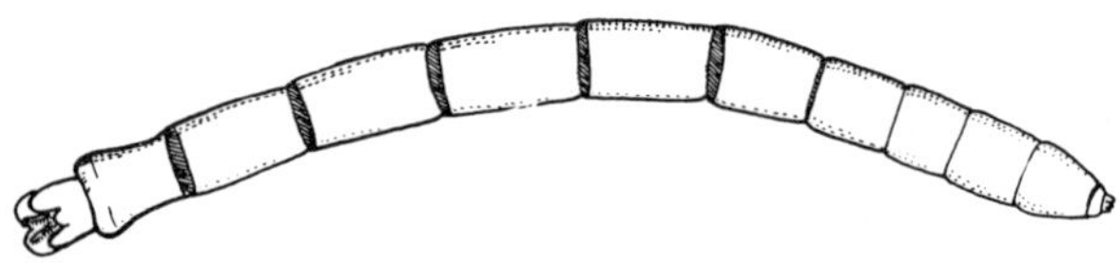

Crane fly larva of *Molophilus* genus.

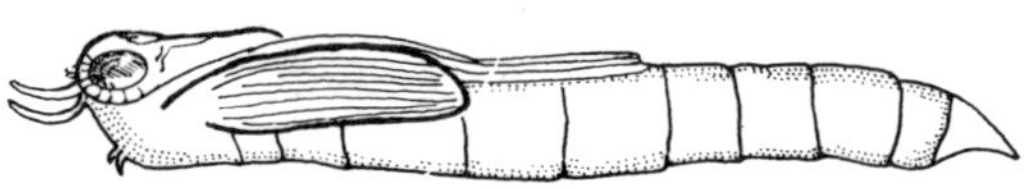

Crane fly pupa of *Molophilus* genus.

Leptotarsus imperatoria. Body length 30 mm.

Australia has 850 described species of crane flies, ranging from some with only 6 to 8mm wing-span to the giant yellow and black species *Leptotarsus imperatoria* which is one of the world's largest with a wing-span of up to 75mm.

Crane flies in the genus *Elephantomyia* have a long mosquito-like proboscis. Others have strangely-shaped antennae. In other groups, females may be wingless or flightless.

Because their flight is slow and clumsy, crane flies are easily caught, but their slender bodies and brittle legs make it vital that a specimen be carefully pinned the moment it is caught. When pinned and dried, specimens are even more fragile and should never be handled.

MOTH FLIES or **OWL MIDGES** (Family PSYCHODIDAE)

These are minute, hairy moth-like flies which frequent damp places. They rarely exceed a wing-span of more than 8mm but they are sometimes seen early in summer walking on a window and spreading and folding their wings as they move. The wings at rest lie like a roof over the abdomen and this enhances the moth-like appearance.

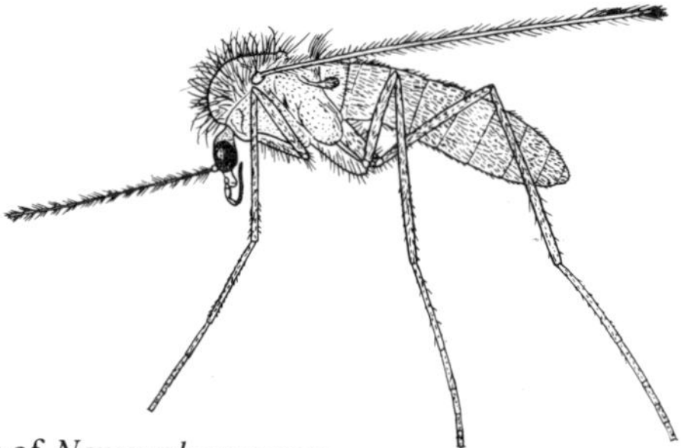

Moth fly of *Nemopalpus* genus.

Larvae are aquatic (see p. 234) and adults rarely live more than a day or two. Overseas species are sometimes called 'sand flies' (genus *Phlebotomus*). These suck human blood and may transmit certain skin ailments and other diseases. Australian species are not known to carry any disease. One species of the blood-sucking type is *P. queenslandi* found in northern Queensland.

MOSQUITOES (Family CULICIDAE)

With 228 species of mosquitoes described in Australia to date, these insects are well represented in the region and have considerable nuisance and economic effects on humans and some animals.

Most species have long piercing-sucking mouthparts, although only the females are bloodsuckers. Males are thinner than females and have feathery, bushy antennae. With many species, bloodsucking is confined to particular hosts or even one host and certain species may feed exclusively on birds, or reptiles, or fish. Males feed on nectar or sap by piercing and sucking.

Mating between mosquitoes takes place in flight and larvae and pupae are always aquatic (see p. 235).

Australia's disease-transmitting species include the Australian malaria mosquito *(Anopheles farauti)*, the brown house mosquito *(Culex fatigans)* which transmits filariasis, and the dengue mosquito *(Aedes aegypti)*. Most discomfort

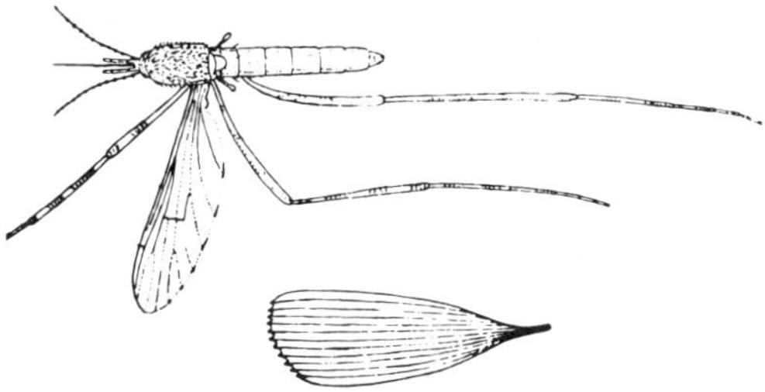

The 'Hexham grey', largest of Australian mosquitoes. Length 7 mm. Also, enlarged scale from wing.

to humans is caused by the brown house mosquito, the common banded mosquito *(Culex annulirostris)* and the salt marsh mosquito *(Aedes vigilax)*, which breeds in brackish coastal waters.

The adults of most mosquito species are active at night, although many move about during the day in shaded, damp locations.

One of the most interesting Australian species is the salt-water mosquito *(Aedes australis)* which breeds in salt water, concentrated by evaporation in rock pools. Others will breed in any place where water is collected—from old tree stumps to hoof marks left in clay.

The largest species, found on Australia's east coast and in the Murrumbidgee Irrigation Area, is the Scotch grey, or Hexham grey *(Aedes alternans)* which has striped legs.

Australian mosquitoes are divided into several groups, The Culicini tribe may be identified by the characteristic horizontal position adopted when feeding, the proboscis and body being held in a straight line.

Those in the Anophelini tribe, including species in the genus *Anopheles*, rest and feed with the body held at an angle to the surface, an action which makes them conspicuous, especially when at rest on a vertical wall.

The common Australian anopheline mosquito *Anopheles annulipes*, found from the far north through to Victoria, is said to have the widest distribution of any Australian species.

TRUE MIDGES, GNATS or **HARLEQUIN FLIES** (Family CHIRONOMIDAE)

These insects resemble mosquitoes but are smaller, ranging in size from a maximum wing-span of 7mm to less than 1mm. The mouthparts of adults are generally not adapted for piercing skin. At rest, true midges raise the forelegs while mosquitoes always raise their hind-legs from the surface.

They will often gather in swarms in the early evening, either near water or around electric light bulbs.

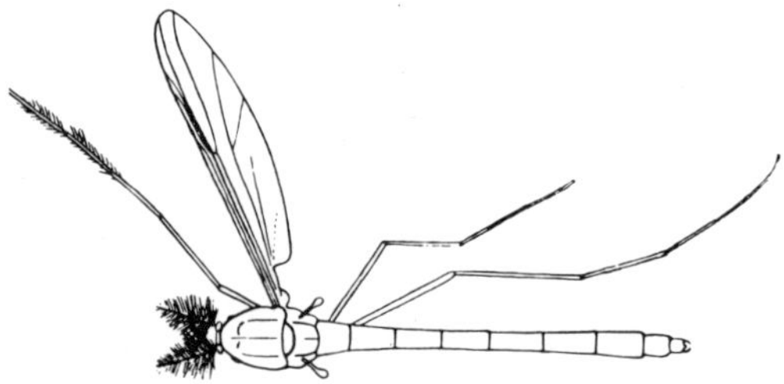

True midge. Length 7 mm.

To date 160 species have been described in Australia, including several in the genera *Clunio* and *Telmatogeton* which occur on rocks by the sea. *Clunio* females are wingless. There is also one submarine genus *(Pontomyia)* which also has wingless females. Males in this genus have their wings reduced to paddles. Nearly all species have aquatic larval and pupal stages (see p. 236) with the larvae being known as 'bloodworms'.

BITING MIDGES (Family CERATOPOGONIDAE)

Often erroneously called 'sand flies', the common name applied to the black flies in the Simuliidae family (see below), biting midges are tiny bloodsucking flies. All have long piercing mouthparts and many are pests of man and other animals, their bite causing a painful allergic reaction which in horses may cause dermatitis. Biting midges are found in the sandy mud of estuaries, sea beaches and the banks of streams. In places their numbers have been known to drive people from sea beaches.

The main Australian species are in the genus *Culicoides*, but 200 species of various genera have already been described. Larvae live mainly in mud or debris, in rotting vegetation, and in salty or brackish waters. They are usually less than 10mm long with no prolegs, or prolegs only on the last abdominal segment. They have retractile gill filaments on the last abdominal segment. The pupal stage is short-lived and relatively inactive.

BLACK FLIES (Family SIMULIIDAE)

Sometimes known as 'sand flies' these small black or dark coloured insects are stoutly built with broad wings and are often bloodsuckers. The Dawson River black fly *(Austrosimulium pestilens)* can make life impossible for man and stock in Queensland and the species *A. bancrofti* is also a widespread pest of livestock.

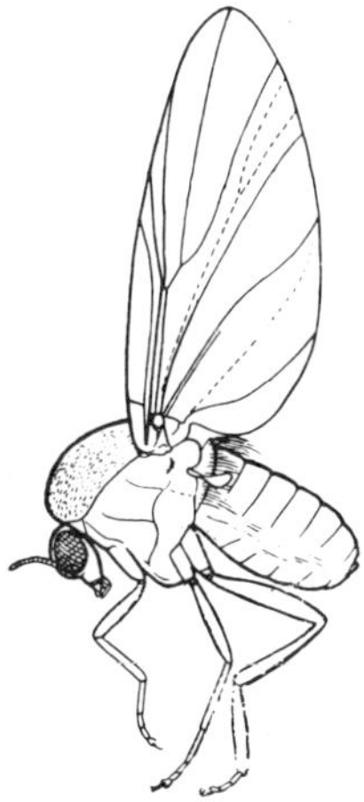

Female black fly. Length 5 mm.

Larvae and pupae are aquatic and larvae have a single sucker at the tail by which they attach themselves to submerged objects. Pupae are identified by two groups of breathing tubes which project from the pupa casing.

NET-VEINED MIDGES (Family BLEPHARICERIDAE)

Found near mountain streams, these small insects have broad iridescent wings which have a fine network of lines, and the long legs and shortish body of a mosquito. Mouthparts are lengthened and adapted to a predatory life with females having a long pointed mandible. Both larvae and pupae are aquatic (see p. 236).

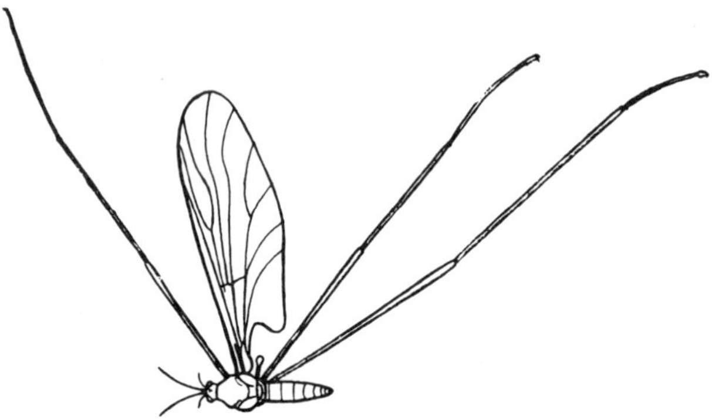

Net-veined midge. Wingspan 24 mm.

GALL MIDGES or **GNATS** (Family CECIDOMYIIDAE)

Gall midges cause small galls to form on the foliage of trees and plants. Most adults are very small, and have hairy wings. Adults of some species are able to land with impunity on spiders' webs.

Galls are caused by the larvae, and one particular yellow-brown hemispherical gall found on the tips of cypress pine shoots results from the presence of the species *Diplosis frenelae* which looks just like the young soft seed vessels of the tree. The larvae lie imprisoned in the gall until it dries and splits whereupon the larvae drop to the ground and pupate.

Distorted seed pods on some wattles are caused by larvae of the species *Cecidomyia acaciaelongifoliae.*

The only pest species of the 111 gall gnats described in Australia is the sorghum midge *(Contarinia sorghicola).* There are also beneficial species including *Zeuxidiplosis giardi* which has been introduced and controls the weed St John's wort.

FUNGUS GNATS or **GLOW-WORMS** (Family MYCETOPHILIDAE)

Found in damp and well-shaded locations, fungus gnats are mainly inhabitants of wet temperate forests. Larvae are usually found near fungi and the 'glow-worms' found in caves in the Blue Mountains of New South Wales and in the Mole Creek caves of Tasmania are fungus-gnat larvae. They may also be found in deep, well-shaded gullies. The light, produced from enlarged tips of the Malpighian tubes, serves as a lure for prey which these larvae trap in the webs on which they hang.

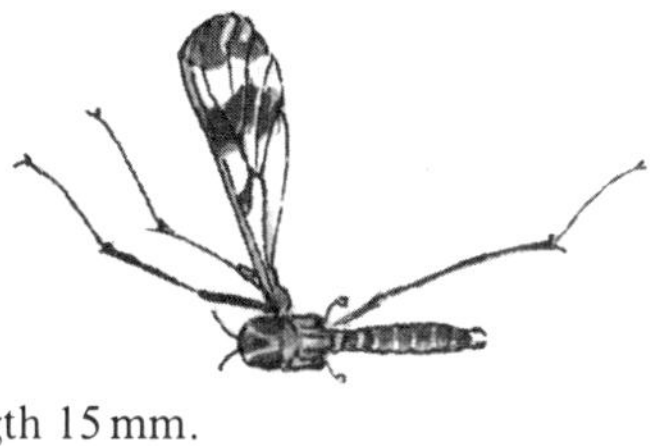

Fungus gnat. Length 15 mm.

MARCH, HORSE or **GAD FLIES** (Family TABANIDAE)

Ranging in length from 5 to 20mm the 240 species of horse flies in Australia are mostly bloodsuckers and can be severe pests of man and livestock. In appearance, these insects have broad heads, a long straight beak and rather broad, hairy bee-shaped bodies.

The bloodsucking Tabaninae subfamily lay their eggs in clusters on shrubs and grass blades overhanging water, or in swamps. Larvae are bullet-shaped with small retractile heads and live in shallow muddy locations in ponds and lakes.

Larvae have rudimentary prolegs on the first seven abdominal segments, and a short breathing tube on the eighth abdominal segment. They may be white, yellow, brown or green. Pupae are said to pass through this stage within the soil.

The orange and black species *(Pelecorrhynchus fulvus)* is grouped in the Pangoniinae subfamily, as is the large black-bodied species with a white spotted abdomen *(Erephopsis guttata)*, from the eastern coastal forests.

Erephopsis guttata. Length 25 mm.

SOLDIER FLIES (Family STRATIOMYIDAE)

When at rest, one wing is folded above the other on the soldier fly's severely flattened abdomen. The segmented antennae end in a slender bristle. At least one genus *(Boreoides)* has wingless females.

Some members of this family might be mistaken for wasps. Many are found near mountain streams and coastal swamps and larvae may be aquatic, or found in moist soil. Larvae of the pasture soldier fly *(Inopus rubriceps)* is a serious pest of pastures and sugar cane in Queensland.

Wingless *Boreoides* female. Length 20 mm.

WINDOW FLIES (Family SCENOPINIDAE)

As their name suggests, these small dark flies may often be found on window panes or other expanses of glass. Their larvae are found in leaf mould and also in the nests of some birds. The larvae of some species are predators of those of other insects.

ROBBER or **ASSASSIN FLIES** (Family ASILIDAE)

With wing-spans ranging from 2 to 75mm, these are slender flies with a humped thorax, long abdomen and a long beak adapted for piercing and sucking. The eyes are prominent and separated, and the face is covered with long hairs. The legs are stout and hairy.

Larvae are cylindrical grubs, tapered at the ends and living in earth or timber. Adults are found in open forests and attack most insects in flight. Prey is held with strong legs and enzymes are injected into it to reduce the flesh to a liquid which the robber fly then sucks.

Australia's largest flies are included in this group, including the giant species *Blepharotes coriarius* which has a wing-span of up to 75mm and is the largest Diptera in Australia. This is a black fly with a bright orange-red upper surface of the abdomen and tufts of black and white hairs along its sides. It is found inland, especially in the Riverina of New South Wales, where it rests on fences and trees.

A giant assassin fly, *Phellus glaucus*, female. Length 55 mm.

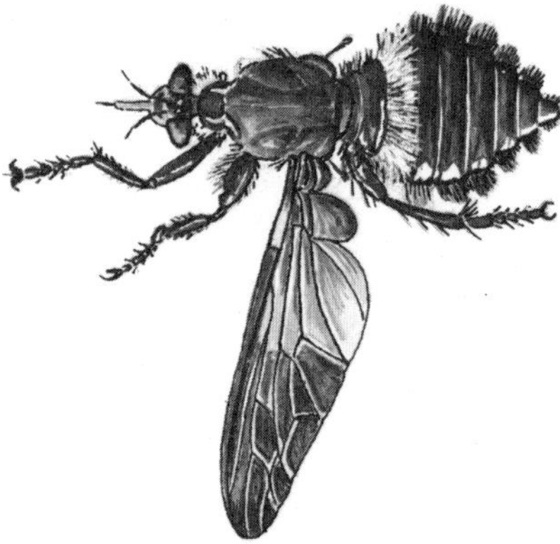

Member of another giant assassin genus, *Blepharotes*. Length 40 mm.

When disturbed, it flies with a loud 'buzz', often carrying prey impaled on its beak. The black and yellow species *Chrysopogon crabroniformis* mimics the large mason wasp (*Abispa ephippium*).

When collected, care should be taken in handling robber flies as they will attempt to impale human flesh with as much alacrity as they do their normal prey.

BEE FLIES (Family BOMBYLIIDAE)

These fat flies with furry bodies have large round heads and slender legs which have no bristles. The mouthparts are formed into a long slender beak and some look very much like bees. Wings have a dark band on the leading edge and two veins always curl towards the tip. Wing-spans of the 400 Australian species range from 70mm to some with a wing-span of no more than 3mm.

Most bee flies are commonly seen above or on blossoms or low above the ground. They are able to dart rapidly sideways or backwards in flight.

It is believed that eggs are placed in the eggs or nests of other insects and that larvae are parasites, possibly of burrowing bees and wasps and also of grasshoppers and locusts.

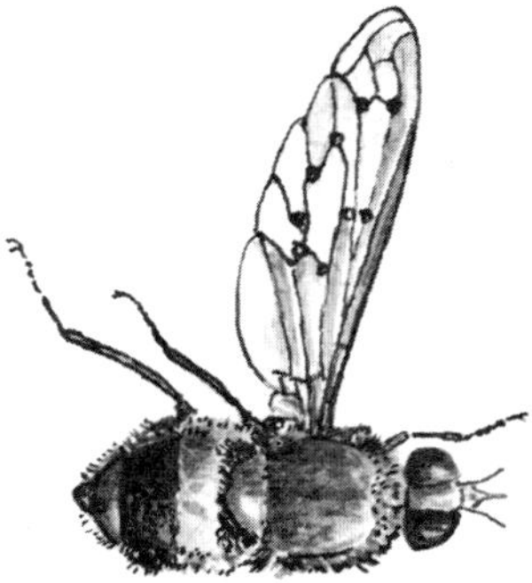

Bee fly of *Hyperalonia* genus. Wingspan 55 mm.

WATER CRUISERS (Family EMPIDIDAE)

Although only about 90 species are known from Australia, water cruisers vary widely in appearance. Generally the head is small and rounded with a slender piercing beak. The thorax is large and the abdomen and legs are slender. It is thought that adults are predators. Larvae breed in decaying vegetable matter or in vegetable debris in streams and ponds.

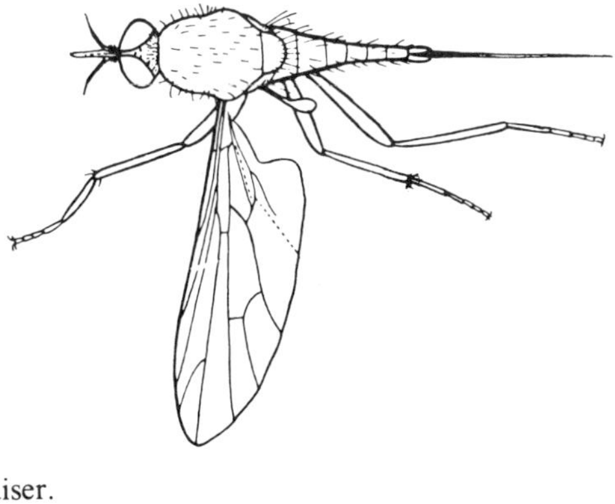

Water cruiser.

In some species there is an elaborate mating ritual with the male flying up and down in the air and carrying either the body of an insect wrapped in a ball of white silky material or sometimes just a ball of white silk. In the former instance, food is presented to the female before mating and she feeds during copulation. Males have also been seen carrying bright petals or a bubble of frothy secretion during the mating dance.

Water cruisers, as their name suggests, may often be found flying over a stream or pond, sometimes in regular formations.

STILT FLIES (Family DOLICHOPODIDAE)

Small flies, rarely having a wing-span exceeding 15mm, they are slender, green or bronze with large rounded heads and thin legs. Mouthparts are rarely developed into a beak.

Adults are seen on tree foliage, or near water. Some are water 'striders' and may be seen in the sun but others seem confined to forests in humid locations.

All stilt flies are thought to be predacious and some may be useful in attacking aphids. Larvae are similar to those of water cruisers, and have been found in mud and beneath bark.

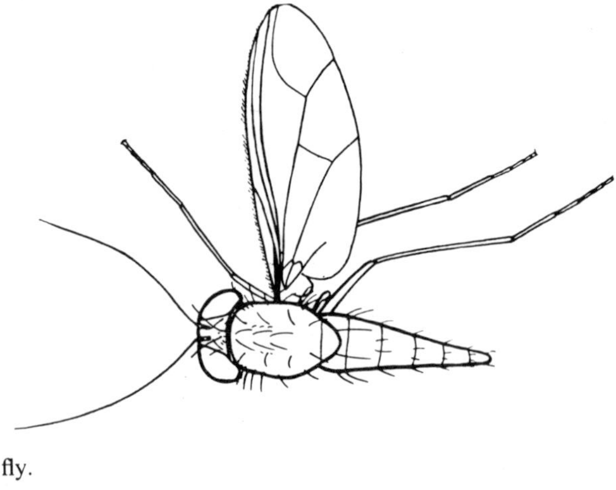

Stilt fly.

HOVER FLIES (Family SYRPHIDAE)

Common, especially in the spring months when they are frequently seen hovering above or resting on flowers, these insects seem similar to bees or wasps with dark brown bodies banded with yellow. Their most distinguishing feature is their ability to hover almost motionless, with their wings vibrating so quickly that they are only a blur. If approached, they dart away and resume their hovering some distance from the original spot.

The larvae of the 168 species known in Australia are mostly terrestrial but conspicuous aquatic larvae are the 'rat-tailed maggots' (see p. 237).

The species *Syrphus viridiceps* is a thin, yellow-banded fly which deposits its eggs singly among aphids. The larva is pale green and has sharp mouthparts which it uses to impale aphids. Once it has captured its prey, the larva stands upright on its hind prolegs and in this position extracts the body fluids of its prey.

Larvae and pupae of the species *Microdon variegatum* often live in ants' nests, picking up food where they can find it.

Wasp mimics are found in the genera *Ceriana* and *Baccha*. Among the bee mimics, another species flies with the bee it mimics. The introduced drone fly *(Eristalis tenax)* is very similar to the hive or honey bee *(Apis mellifera)*.

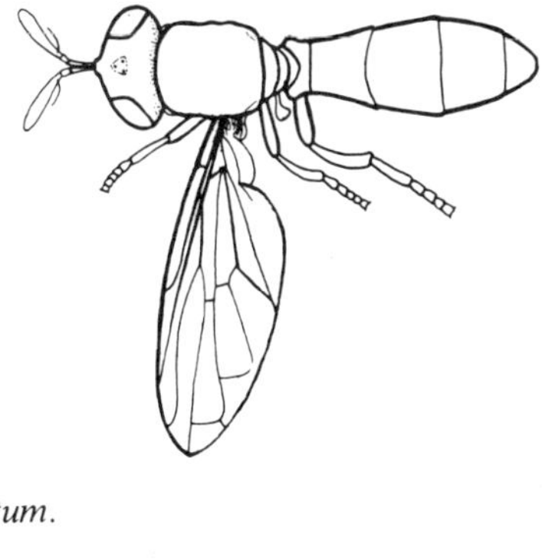

Microdon variegatum.

WASP FLIES (Family CONOPIDAE)

In some ways very similar in appearance to the hover flies (see above) the wasp flies may often mimic wasps, but differ from the hover flies by having a ptiline fissure present above the bases of the antennae.

Wasp flies have a narrow 'waist' on the abdomen and are black with yellow bands. Australia has at least 75 known species and most have well developed organs at the end of their abdomen. However, unlike hover flies, they are slow fliers and may often be caught on blossoms. Their larvae are parasites of wasps and bees.

AUSTRALIAN STALK-EYED FLIES (Family PLATYSTOMATIDAE)

Although the true stalk-eyed flies of overseas (family Dopsidae) are not represented in Australia, the three species of the genus *Achias* all have eyes on 'stalks' or lateral extensions of the head. This phenomenon is found chiefly in the males, the females' stalk being not nearly so long or conspicuous. They are found only in North Queensland and New Guinea and little is known of their life history.

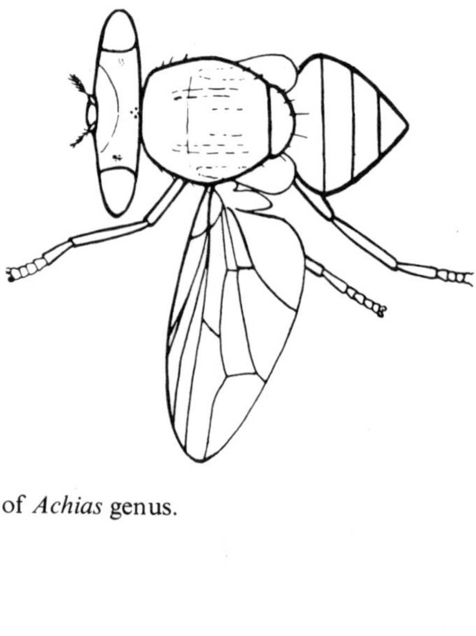

Stalk-eyed fly of *Achias* genus.

FRUIT FLIES (Family TEPHRITIDAE)

On most roads leading into the state of Victoria, 'fruit fly inspection posts' have been set up. The aim of these inspection points is to prevent the entry of fruit which might contain the larvae of the Mediterranean fruit fly *(Ceratitis capitata)* and the Queensland fruit fly *(Dacus [Strumeta] tryoni)* into Victoria. These small insects, with a wing-span of 13mm or less, attack a wide variety of cultivated and native fruits. The related cucumber fly *(D. cucumis)* will also attack cucumbers and marrows. Fruit flies are usually brightly coloured and often have banded or mottled wings and a strongly-pointed abdomen. Frequently these flies carry their wings in a drooping position while moving about foliage.

Damage is caused by the female depositing her eggs inside the flesh of fruit. The eggs hatch into waxy maggots which have a tapered head armed with hook-like mandibles. There is a breathing tube at the tail. The larvae feed on the fruit flesh and reduce it to a rotting mass. After the fruit falls, the larvae pupate in the soil, enclosed in the final larval skin.

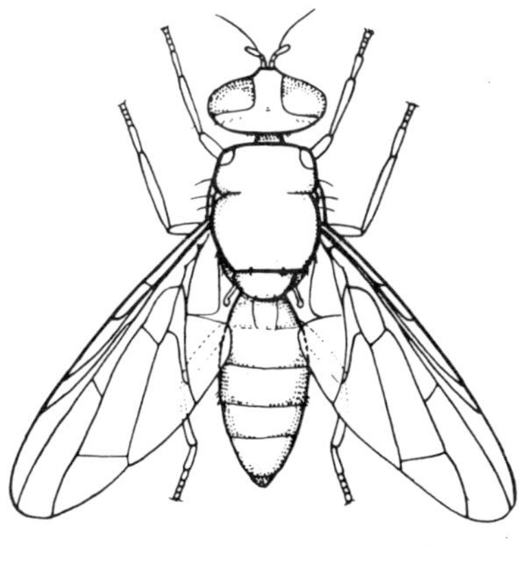

Queensland fruit fly.

The Mediterranean fruit fly has a body banded with brown and yellow; the Queensland fruit fly is reddish and similar to a wasp with unmarked wings.

The species *Termitorioxa termitoxena* live in termite nests in trees in the Northern Territory. Other species live in the seeds of certain burrs (e.g. the Bathurst burr seed fly, although this is said to have little effect in reducing the extent of this plant). A Mexican species, *Procecidochares utilis*, was imported into Queensland to control crofton weed. Eggs are laid in young shoots which become galled by the larvae and the plants fail to develop.

KELP FLIES (Family COELOPIDAE)

Large swarms of these flies are often found on sea beaches where they swarm over seaweed cast ashore by storms, or near rocks in sheltered locations. The larvae live on seaweed and most adults seem to remain near the beaches. Kelp flies seem to be confined mainly to the south of the continent and the most common species is *Chaetocoelopa sydneyensis*. They are medium-sized and yellowish in colour. The larvae are one of the few kinds of insects which can tolerate salt.

TOMATO FLIES (Family LONCHAEIDAE formerly Sapromyzidae)

The green tomato fly *(Lamprolonchaea brouniana)* has a metallic sheen. It deposits its eggs in damaged or rotting fruit and potatoes. Maggots are small and yellow.

LEAF MINERS (Family AGROMYZIDAE)

The adult is very small, and these flies are perhaps better known for the damage caused by their larvae. Larvae of the species *Agromyza phaseoli* tunnel into the stems of bean plants and cause them to wilt and die. More usefully, larvae of the lantana seed fly *(Ophiomyia lantanae)* live in lantana seeds and reduce the quantity of viable seeds from this pestilential plant. Yet another species excavates tunnels and pupates in the leaves of sow thistles.

Leaf miner.

VINEGAR or **FERMENT FLIES** (Family DROSOPHILIDAE)

These small, yellowish flies are probably best known for the use which has been made of them in laboratories rather than for the native species whose larvae live mainly on decaying fruit or fungus. *Drosophila* have been widely used in experimental genetics and are of particular value because of the number of generations which can be produced under optimum laboratory conditions. This allows scientists to obtain results in far less time than would be possible with most other forms of animal. One species, *D. funebris*, was said by McKeown to be able to breed in crude formalin used to store zoological specimens in the Australian Museum. Such an environment is deadly to most other forms of animal life.

HOUSE FLIES (Family MUSCIDAE)

With some 200 species of these flies known from Australia, the family includes two major pests. They are the common house fly *(Musca domestica)* which is recorded as a carrier of various bacteria, and the pestilential bush fly *(Musca vetustissima)* which is a constant annoyance to man and his animals in the summer months.

Larvae of the house fly live in refuse and excrement and the flies may carry numerous germs on their hairy bodies. Females lay some 120 eggs which hatch into yellow waxy maggots.

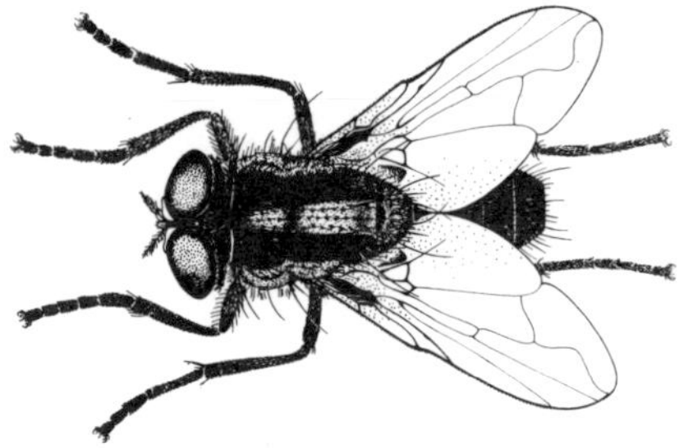

Bush fly.

Larvae of the genus *Passeromyia* live in birds' nests.

The stablefly *(Stomoxys calcitrans)* is a bloodsucker of man and domestic animals, while the buffalo fly *(Haemotobia exigua)* has become a pest of stock in northern Australia.

BLOWFLIES and **BLUEBOTTLES** (Family CALLIPHORIDAE)

These large flies are known for both their size and for the buzzing noise they make. Blowflies are not dangerous disease-carriers, though their numbers can make them a household nuisance. While some blowflies will feed mainly on nectar and other sweet liquids, it is their egg-laying habits which make them pests. In summer, meat cannot be left uncovered without the risk of it being 'fly blown' and in a short while having a large population of maggots.

The species of greatest economic importance is the metallic-green Australian sheep blowfly *(Lucilia cuprina)*, which lays its eggs on damp parts of a sheep's body; the maggots then eat into the flesh of the animal. After a sheep has been affected by a 'primary blowfly strike' it may be attacked again by secondary species. These include the hairy maggot blowfly *(Chrysomya rufifacies)*, whose larvae both compete with the primary maggots and also prey upon them. Some small blowflies may be metallic green or blue in their adult form and are also known as 'bluebottles'. Many species do

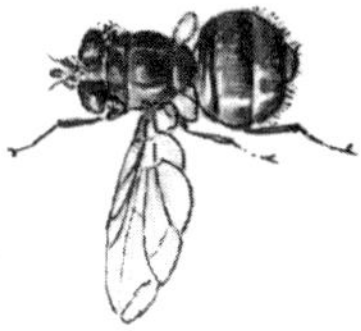

Common blowfly. Length 13 mm.

not affect man and their larvae parasitise earthworms and snails. Others live in the nests of ants and termites. To date, 135 species of blowflies are known in Australia.

FLESH FLIES (Family SARCOPHAGIDAE)

With their plumed antennae, and grey bristly bodies with stripes running the length of the thorax, flesh flies are usually seen in carrion and dung. Larvae are usually deposited on decaying organic matter and play an important part in reducing it to substances which can be absorbed by the soil.

Some species are parasitic, the larvae living in wasps' nests. Others are said to parasitise the poisonous funnel-web spider *(Atrax robustus)* and others parasitise plague locusts.

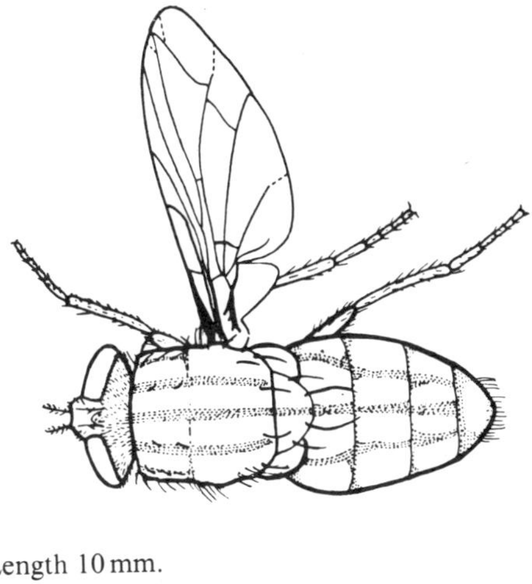

Flesh fly. Length 10 mm.

TACHINID FLIES (Family TACHINIDAE)

With some 520 species described from Australia, these blowflies with their strong bristly bodies range considerably in size. While some are only half as big as a house fly, others may have a 35mm wing-span. Larvae are parasites of many types of pest insects and are therefore beneficial to man.

Tachinid fly of *Euamphibolia* genus. Length 22 mm.

BOT or **WARBLE FLIES** (Family OESTRIDAE)

These medium to large hairy insects have a bee-like appearance, with eyes proportionally smaller than those of most other flies. Overseas, the flies deposit eggs on the fur of an animal, close to its mouth. By licking, the animal swallows the larvae, and they eventually reach the stomach where they anchor themselves to the walls. The introduced sheep bot fly *(Oestrus ovis)* is present but is said not to be an important problem. The native species *Tracheomyia macropi* has larvae which infest the trachea of red kangaroos. While bot flies are a significant pest in some overseas countries, they are not considered a serious problem in Australia.

LOUSE FLIES or **SHEEP KEDS** (Family HIPPOBOSCIDAE)

These bloodsucking flies have evolved as parasites which live in the hair of mammals and the feathers of birds. They have flattened bodies, extended mouthparts and strong legs, which gives them a superficial similarity to lice or ticks, although louse flies have only three pairs of legs.

In most species, the wings are developed. However, the introduced and spider-like sheep ked or 'tick' *(Melophagus ovinus)* has only vestigial wings. In cooler parts of Australia, severe infestation can cause economic losses, either through blood-stained wool or by weakening animals through anaemia.

Louse flies retain larvae within their body until they are ready to pupate and the puparium is often attached to the hair or skin of the host. Various species infest Australian native marsupials and birds.

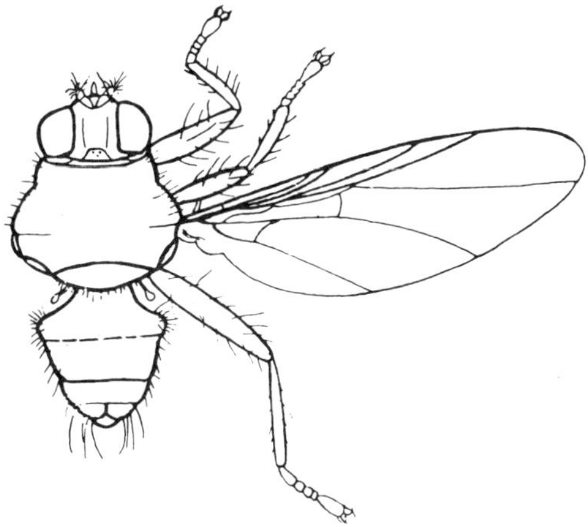

Ornithomyia australasiae, a parasite of birdlife. Length 6 mm.

CADDIS FLIES

Order TRICHOPTERA (hairy-winged)

Related to Lepidoptera (see below), caddis flies are like small, drab-coloured moths. Their wings, however, are covered with thick silky hairs, as are their bodies, and their antennae are long and slender. When caught and handled, the hairs of a caddis fly come away less easily than the scales on a moth or butterfly. At rest, the wings are held like a roof over the body.

Adults fly mainly at dusk and during the night, although they may be attracted to lights. They live mainly near running streams or lakes and ponds and are more common in the colder parts of Australia. In the daytime, they may be found resting on tree trunks on the banks of streams and lakes.

Some 260 species of caddis flies have been described from Australia.

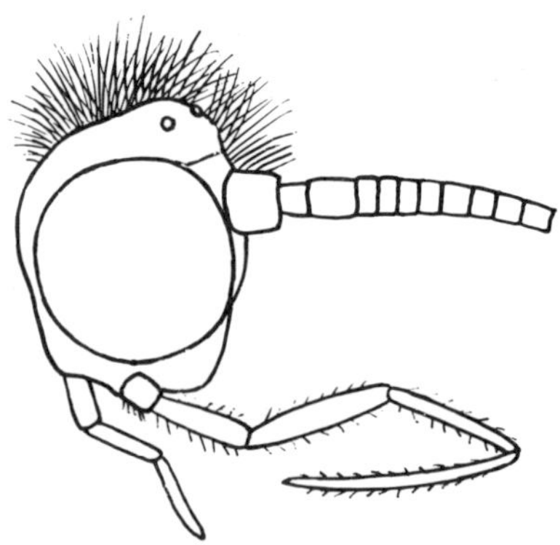

Head of caddis fly of *Hydrobiosis* genus.

One of the most interesting Australian species is the world's only marine caddis fly *(Philanisus plebeius)*, which breeds in rock pools in southern Australia and New Zealand. Another species, *Symphitoneuria wheeleri*, breeds in lakes which may contain a higher salt content than the sea.

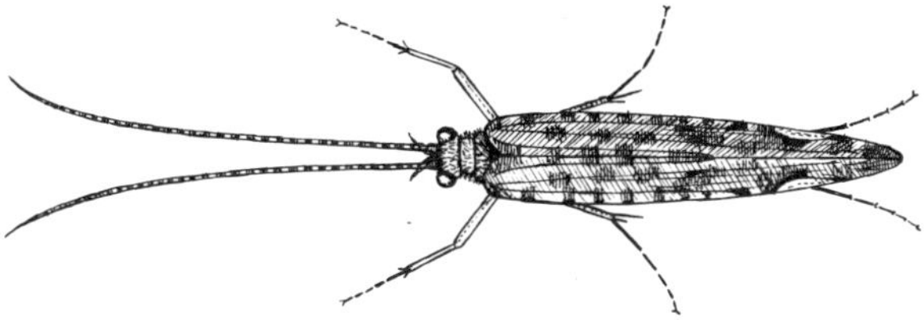

The marine caddis fly, *Philanisus plebeius*. Length 8 mm.

While adult caddis flies are not well known, their larvae and pupae are more commonly found in streams and lakes in cooler regions and look like a sort of 'underwater case moth' (see pp. 237–9). Many groups are more easily identified by the distinctive form of their larval cases than by the adults.

One of the best known day-flying caddis flies is the misnamed Shannon 'moth' *(Asmicridea grisea)*. This species is abundant in parts of Tasmania and when the pupae are transformed into adults, large numbers of fish, and particularly trout, concentrate to feed on the pupae and emerging adults.

Adult caddis flies may range in size up to 40mm in length. Details of eggs and immature stages are described on pp. 237–9.

MOTHS AND BUTTERFLIES

Order LEPIDOPTERA (scale-winged)

Insects in this order possess wings which are covered by regular overlapping rows of scales. Mostly, their mouthparts or proboscis consist of a sucking tube *(haustellum)*, which coils like a watchspring when not in use. However, in some moths the mouthparts are vestigial and the adult is unable to eat after it has emerged from pupation.

In Australia, 11 000 species have been described and

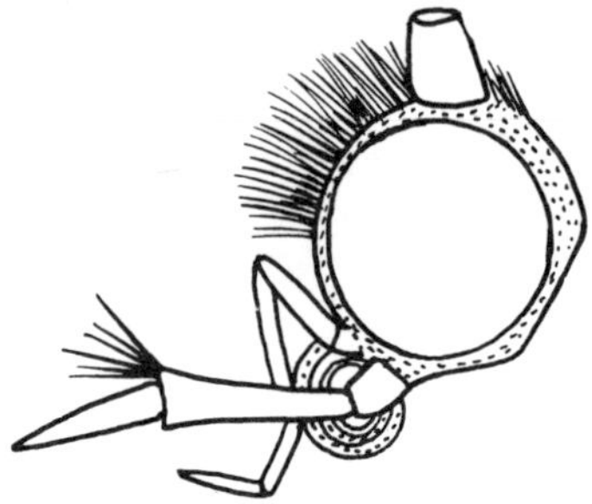

Head of clothes moth, antenna truncated.

possibly a further 50 per cent have yet to be named. This makes Lepidoptera the order with the second-highest number of species in Australia, a feature which parallels world trends.

All moths and butterflies have compound eyes. Sometimes a pair of simple eyes *(ocelli)* are also present above the compound eyes.

All Lepidoptera undergo complete metamorphosis and from eggs laid on a host plant, larvae (caterpillars or grubs) are hatched. These usually have a hardened head and a pair of jaws *(mandibles)*. Three pairs of true legs occur at the front and there are anal claspers at the end of the body. Between the true legs and anal claspers there may be varying numbers of false or prolegs. Larvae may be hairy or smooth-skinned, and may have unusual fleshy spines or tubercles.

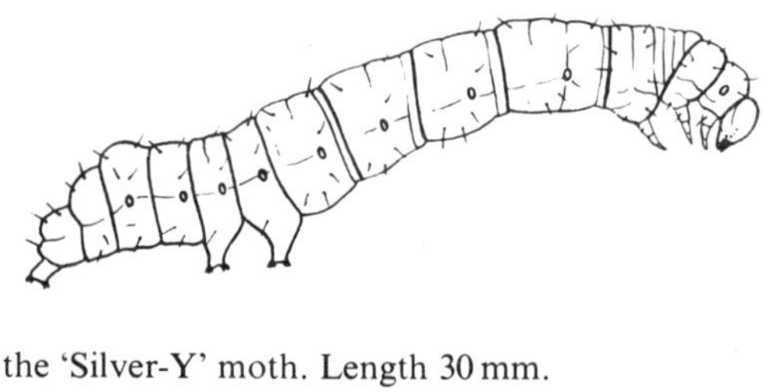

Larva of the 'Silver-Y' moth. Length 30 mm.

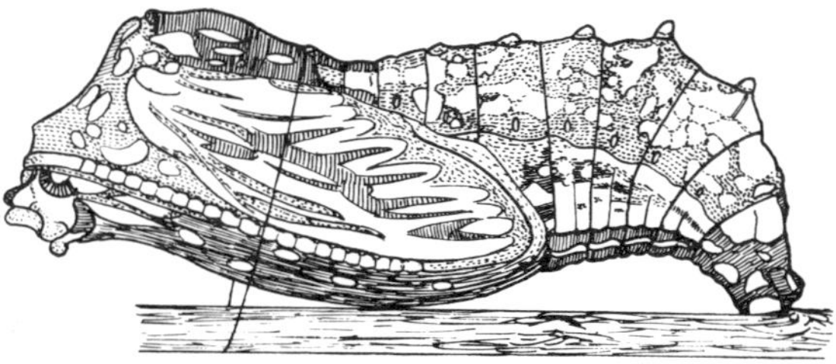

Swallowtail butterfly pupa with silk girdle. Length 27 mm.

In time the larvae pupate, usually spinning a case except in butterflies where the pupa is naked. When the imago or new adult butterfly emerges from pupation, the wings have to expand by having blood circulated through them to stiffen the membranes. After expansion, the wings must then dry and harden before the adult can fly. Some moths have reduced wings (or none at all) and do not fly. Adults may live from a few days to several months.

While some moths and butterflies are dull in colour and rely on their camouflage *(cryptic coloration)* to protect them, others have bright markings and may in turn be distasteful to their enemies. Others may mimic bees or wasps. Similarly caterpillars are often well camouflaged and may even rest in the likeness of a twig. Others need no camouflage as their hairs are poisonous or irritating both to man and other natural enemies.

The huge birdwing butterflies of New Guinea and the rain forests of North Queensland are among the largest in the world and may have a wing-span exceeding 190mm. The first New Guinea birdwing was obtained by Rothschild who 'captured' it by shooting it down with a shotgun. The giant atlas moths of the same region are even larger and may have a wing-span of 250mm or even more. In contrast, the tiny leaf miner lepidopteran may never attain a wing-span of more than 3mm.

Until the 1950s, the viewpoint of the layman had been followed by scientists who divided the Lepidoptera into two broad divisions. These were Heterocera (other antennae) for moths and Rhopalocera (clubbed antennae) for butterflies. This was later changed to Heteroneura (different venation on front and back wings) for moths and Homoneura (similar venation) for butterflies.

However, by 1958 such distinctions had been discarded. By the latest classification the moths and butterflies of Australia are placed in four suborders which in turn encompass 23 superfamilies and 75 families.

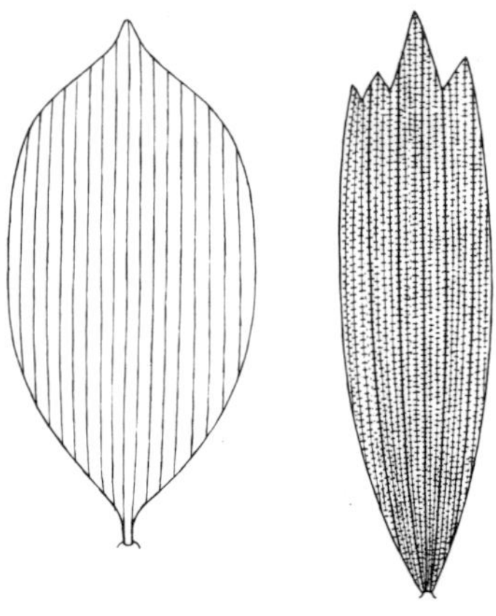

Magnified scales from Lepidoptera wings.

Only two of these superfamilies—the Hesperioidea and the Papilionoidea—represent butterflies.

While the general order of Common will be followed here, butterflies are set apart at the end of the section on moths for the sake of clarity. Butterflies are here referred to by families and except where no common name exists, will be listed by their traditional group names.

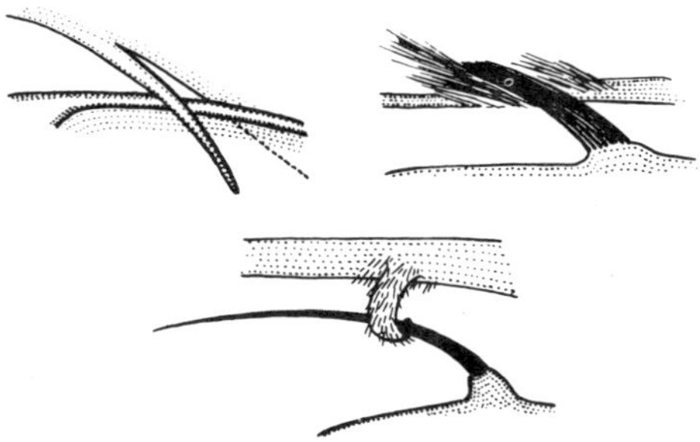

Examples of Lepidoptera wing coupling mechanisms.

Definitions of moths and butterflies used to be:

Moths	Butterflies
Coupling apparatus *(frenulum)* present, holding wings together in flight.	Coupling apparatus absent; function performed by humeral lobe of hind-wings.
Antennae rarely or not clubbed.	Antennae clubbed.
Mostly fly by night	Mostly fly by day.
At rest, wings outspread or wrapped round the body.	At rest, wings raised above the body.

As with all large orders of insects, there are exceptions to each of the listed features. The skipper butterflies (family Hesperiidae) form a link between moths and butterflies and even today, are placed in a separate superfamily. However, on a suborder basis, the factors now used by scientists to group the Lepidoptera include wing venation (not visible unless scales are removed), mouthparts, the legs of adults, the larvae, and genital openings.

Common (1970) includes very detailed keys for adults and larvae of all groups from suborders to families and readers needing greater detail are advised to refer to his work. In the following pages, identification has been kept as simple as possible so that superficial identification can be made from the more obvious features.

Moths

Groups included have been selected on the basis that they are sufficiently well-known and have well-accepted common names.

SWIFT or **GHOST MOTHS** (Family HEPIALIDAE)

About 100 species of these moths are known in Australia; they are rapid fliers with hairy bodies and have rather narrow wings, often with interesting patterns. Caterpillars often 'dig' into the timber of growing trees, cutting channels in the bark or else they burrow in the soil and eat roots. Grubs of swift moths in the genera *Oncopera* and *Oxycanus* dig into the soil and emerge in the dark to eat plants.

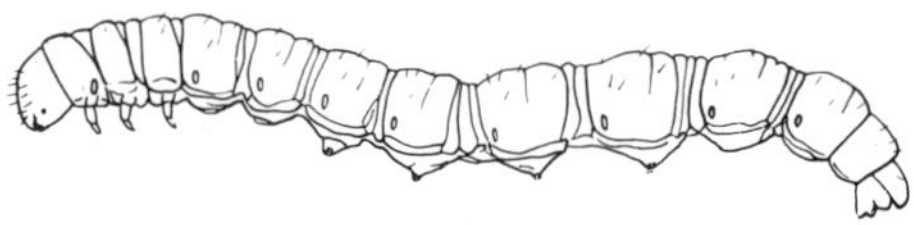

Bent-wing swift moth larva. Length 130 mm.

The bent-wing swift moth *(Zelotypia stacyi)* may grow to a wing-span of 170mm. It is cinnamon-brown with silver markings and raised 'eye spots' near the tips of the front wings. Females have less brilliant colouring, and may grow to 225mm wing-span. The caterpillars attain a length of 90mm and bore in eucalyptus trees. This huge moth is found mainly in the central coastal areas of New South Wales. Swift moths in the genus *Aenetus* may be found in colours ranging from bluish-green and pink in female *A. paradiseus*, to pale blue in male *A. splendens*.

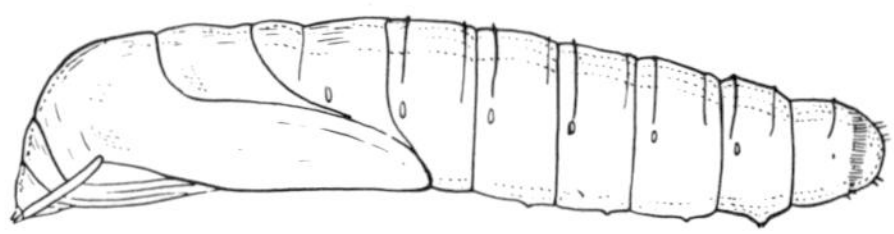

Bent-wing swift moth pupa. Length 110 mm.

Bent-wing swift moth.

The Murray River ghost moth *(Trictena argentata)* is large and grey with wings marked with silver-white bands on the forewings. Its grubs breed in red gums along the rivers and are known as 'barti' or 'bardee' grubs, often used by fishermen as bait. In earlier times, both adults and grubs were eaten by aborigines.

WOOD MOTHS (Family COSSIDAE)

Another family whose grubs are wood borers is the fast flying wood moth which, in its adult form, is usually light grey with a heavy body and narrow, densely-scaled wings, often with attractive patterns in the forewings. The largest Australian species is the giant wood moth *(Xyleutes boisduvali)* whose females may have a wing-span of up to 250mm and an abdomen the size of a small sausage. Its fleshy yellow grubs can attain a length of 150mm and remain in this form within eucalyptus trees for up to five years. A related species is *X. leucomochla*, whose larva forms a silken tunnel in the soil but emerges to feed on the roots of wattle trees in South Australia. This larva is the true 'witchety grub', still relished by some aborigines. The female of the species *X. amphiplecta* is flightless and her short wings serve only as a body cover.

Wood moth, *Xyleutes donovani*. Wingspan 75 mm.

LEAF ROLLERS or **BELL MOTHS** (Family TORTRICIDAE)

Named after their caterpillars, which spin leaves together, the adults usually have wings which are wide and broad; some 800 species have been described from Australia. They include various pest species, including the light-brown apple moth *(Epiphyas postvittana)* whose grub burrows into growing apples and oranges and the codling moth *(Cydia pomonella)*. Smaller species have grubs which burrow into leaves and flowers. The grubs of *Arotrophora* infest banksia in south-eastern Australia.

The species *Epinotia lantana* was introduced and has been effective in reducing the growth of lantana in Queensland and northern New South Wales.

CASE or **BAG MOTHS** (Family PSYCHIDAE)

Better known for their larvae, which are encased in a distinctive case or bag, case moths in Australia have females which may be fully-winged, partly-winged but flightless, or wingless. The adults are small, and when wings are present in females they rarely have a wing-span exceeding 50mm

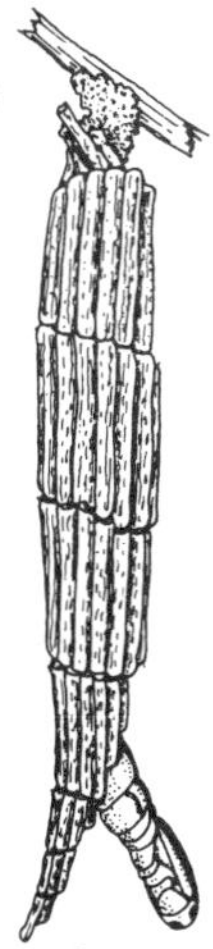

Case moth of *Oeceticus* genus. Wingspan 40 mm.
Right: Case.

and are often much smaller. The wings are usually slender and bodies are extensible. The larvae spin a case of silk and to it are attached fragments of twigs, sand, moss, lichen, bark or leaves which are of a distinctive form for each of Australia's 150 species.

The cases made by the larvae have two openings. One is for the head which projects with the strong legs and the other is for the ejection of excrement. The larvae live off host plants and in the eucalyptus-feeding *Clania ignobilis*, the case is surrounded by parallel lengths of twigs. In this and some other species, the females never leave the case and fertilisation by winged male moths takes place within the case, the male extending its abdomen to fertilise the female through the hole at the back of the case.

The female of *Cebysa leucoteles* is bright orange and green. She is winged, but does not fly. Eggs are often laid by the adult female within the case and when they hatch, young emerge through the rear hole and drop by means of silken strands to a twig or something to which they can attach. They soon begin to make their own cases, which are enlarged as the larvae grow.

CLOTHES MOTHS (Family TINEIDAE)

With 140 species described in Australia, these moths rarely attain a wing-span exceeding 25mm and are often much smaller. Most are inconspicuous and may be grey or white. Some larvae make portable cases and some feed on dead vegetable material including fungi. Others are found in the nests of meat ants and may live by scavenging. The species which cause damage to clothes are the introduced members of the genus *Tineola*. The moths lay eggs singly or in groups on fabrics. These hatch in about a week and when fully grown, the grubs, which are no more than 10mm long, spin silken tubes, often with fibres of the material woven into the walls. These grubs may cause considerable damage by feeding on woollen clothing and carpets for a period of some 65 days or more before finally pupating.

Species in the genera *Nemapogon* and *Niditinea* infest grain and other foods.

CLEAR-WING MOTHS (Family AEGERIIDAE)

These are small moths with narrow, partly transparent wings. Adults are small, with a wing-span not exceeding 20mm but bright red and black. Often these moths are similar in appearance to some types of wasp, although their antennae are broader towards the tip. Larvae are borers in branches of trees and shrubs and those of the introduced species *Synanthedon tipuliformis* are pests of currant and gooseberry bushes in south-eastern Australia.

Clear-wing moth.

POTATO MOTHS (Family GELECHIIDAE)

Small and inconspicuous, these moths have larvae which tunnel in seed pods and leaves and tubers. Of the 480 species described in Australia, possibly the best known is the potato moth *(Phthorimaea operculella)*. The adult lays its eggs on the leaves of the plant and the larva enters the leaves and passes down the leaf stalk, causing the main part to die. The larvae also enter the potatoes, both in the ground and in store. Related species include the cotton pest, pink bollworm *(Pectinophora gossypiella)* of the Ord River area and a related species in Queensland. The Angumois grain moth *(Sitotroga cerealella)* is a similar pest of grain.

MANY-PLUMED MOTHS (Family ALUCITIDAE) and **PLUMED MOTHS** (Family PTEROPHORIDAE)

Although growing only to a wing-span of 17mm, the many-plumed moths have each wing deeply cleft into six separate plumes. The plumed moths have half as many plumes on each wing. Larvae of these beautiful and delicate moths feed on *Bignonia* and *Tecoma* plants.

CASTNIID MOTHS (Family CASTNIIDAE)

Superficially resembling butterflies both in wing-shape and with their clubbed antennae, these moths are often yellow and black, though at rest the camouflaged forewings often cover the brighter areas.

They are day fliers, often skimming the ground over grass, and in forests, before coming to rest in patches of sunshine.

Castniid moth of *Synemon* genus. Wingspan 30 mm.

Larvae are soil dwellers, and live off roots of bushes. The main Australian species are included in the genus *Synemon* and one of the largest and most impressive is *S. magnifica* from the Blue Mountains of New South Wales. Similar species are found elsewhere in Australia.

FORESTERS (Family ZYGAENIDAE)

Black, metallic blue, or green, these small moths have broad wings fringed with hair. They rarely have a wing-span exceeding 25mm and are day fliers. Larvae are fat and covered with thick hair; feeding on flower buds.

Forester moth. Wingspan 25 mm.

CUP MOTHS (Family LIMACODIDAE)

The name of this group comes from the cocoon of the pupa which is cup-shaped and has a circular 'lid' over the opening. It may often be found on eucalyptus trees. Brightly coloured grubs or caterpillars of species in the genus *Doratifera* possess hairs which will sting man. Adults have a wing-span up to 42mm and have broad wings and a short, fat, hairy body. Caterpillars of some species can be found on more than one host plant.

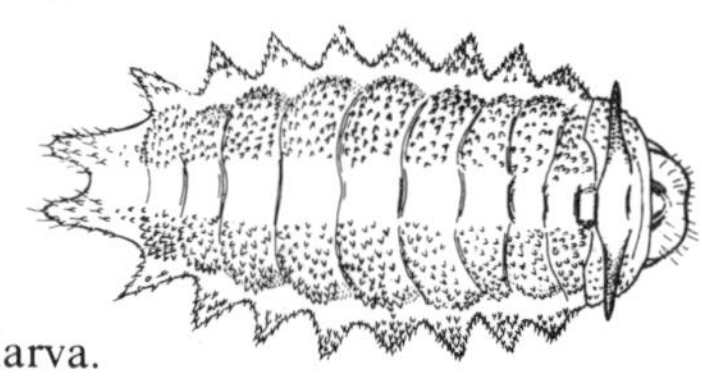

Stinging cup moth larva.

Cup moth of *Doratifera* genus. Wingspan 25 mm.

GRASS MOTHS (Family PYRALIDAE)

These are small moths which have particularly long legs and a beak-like proboscis. At rest, some species curl the abdomen over the back and rest with wings outspread. Others hold their wings above the body. The largest species is *Hypsidia erythropsalis*, which attains a wing-span of 67mm and is found in the rain forests of northern Queensland. However, most species are about 25mm across the wings.

To date 1100 species of grass moths have been described in Australia; the most famous of these is *Cactoblastis cactorum* which was introduced from Argentina. From 2750 imported cactus moth eggs, sufficient moths were hatched to spread through those areas of Queensland over-run by prickly pear and the insect virtually eliminated this pest plant. Larvae bore into prickly pear leaves and cause the plant to die. Other species are pests of rice *(Tipanaea patulella)*, and of sorghum and cultivated fruits *(Dichocrocis punctiferalis)*. However, the larva of *Titanoceros thermoptera* is of benefit to man as it preys on the egg masses of the bag-shelter moth which damages boree and other inland fodder trees.

The species *Nymphula polydectalis* has larvae adapted to an underwater life. They have long breathing tubes or filaments which act as gills; and live in silken tubes spun of submerged water plants or else in small cases which float on the surface. They are found in northern and eastern Australia and are common in Albert Park Lake in Melbourne, Victoria.

HOOK-TIPPED MOTHS (Family DREPANIDAE)

Although only five species are known in Australia, these moths, with their 50mm wing-span, are conspicuous for the extension or hook at the tip of the forewings. They are known from the rain forests of northern Queensland.

Hook-tipped moth. Wingspan 35 mm.

LOOPER CATERPILLARS (Family GEOMETRIDAE)

With no legs at the middle of the body, these caterpillars move by a series of loops, first stretching the body, taking hold with the three pairs of true legs and then looping the body until the two pairs of legs at the rear can grasp the twig or ground surface near the front legs. The scientific name of the family Geometridae means 'earth measurers'—an apt description. To date, 1200 species have been recorded in Australia. Many geometrid adults lie flat on bark and have well camouflaged, dark-patterned wings. Some caterpillars in this family are pests of *Pinus radiata*, Cape weed and wattles.

One of the most interesting of this family is the dead-leaf moth (genera *Circopetes* and *Onychopsis*), which rests on a twig with its abdomen and wings looking like dying leaves.

One of the most brilliant species is *Epidesma queenslandica*, an impressive moth with a 62mm wing-span found in the rain forests of northern Queensland. This moth has black wings with orange and red markings, a hairy red-brown thorax and bright green abdomen. In contrast the females of the genus *Zermizinga* have strangely shaped and abbreviated wings which prevent them from flying. Their larvae feed on the cypress pine.

Many emeralds or green moths are also included in this family and are normally found in tropical rain forests.

Geometrid moth, of the green *Crypsiphora* genus. Wingspan 30 mm.

WHITE-STEMMED GUM MOTHS (Family ANTHELIDAE)

These are large, impressive moths with hairy bodies and large wings. Caterpillars are also covered with dense hairs and bristles and when handled may cause an itching skin rash. The white-stemmed gum moth *(Chelepteryx collesi)* has very large caterpillars which attain a length of up to

The white-stemmed gum moth. Wingspan over 100 mm.

150mm and feed on gum trees. Its cocoon is also covered with stiff bristles, cast by the caterpillar, which can cause itching if they penetrate human skin. Smaller but similar species feed on wattles, pine trees and gums.

EMPEROR MOTHS (Family SATURNIIDAE)

Among Australia's best known moths, the emperors have many claims to fame. The large emperor gum moth *(Antheraea eucalypti)* is known to adults and children throughout eastern Australia as these large moths with a wing-span of up to 125mm are often attracted by house lights. The large blue-green caterpillars are also well-known as are the large, hard, oval cocoons made when the caterpillars pupate.

The emperor gum moth. Wingspan 35 mm.

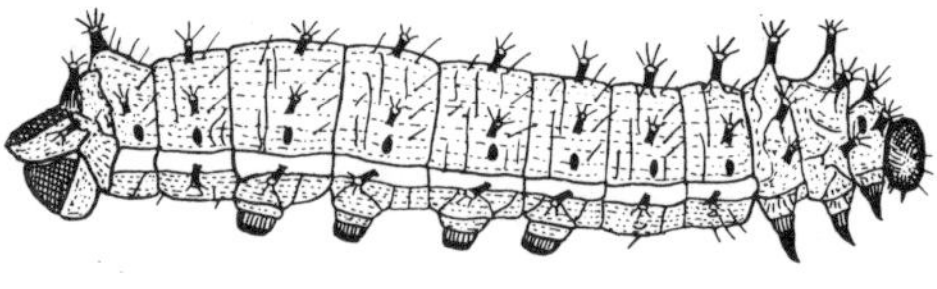

Caterpillar larva of the emperor gum moth. Length 75 mm.

Less well-known, as they are found mainly in the rain forests of northern Queensland and in New Guinea are the huge Hercules moths *(Coscinocera hercules)* which are the largest lepidoptera in the world, some having a wing-span of from 250 to 350mm. The male Hercules moths also have long tails from the rear wings which make them so large that they would easily cover a plate 250mm in diameter. These moths all have conspicuous wing 'eye spots' and large feathery antennae and furry bodies. Some 13 species are known in Australia.

HAWK MOTHS (Family SPHINGIDAE)

With their narrow pointed forewings, short hind-wings and a tapered body, often twice as long as the width of the wings, hawk moths are fast fliers and may often be seen at dusk, hovering over the flowers whose nectar they eat with their long curled proboscis. Feeding takes place while the moth is hovering.

The double-headed hawk moth *(Coequosa triangularis)* is named after its large, bright green caterpillar, which may grow up to 150mm long, and has anal claspers or false legs larger than its head. Also it has two huge eye spots towards the end of its body, giving it the impression of having two heads. The adult double-headed hawk moth has a wing-span of 200mm and has rich brown forewings marked with a darker area and yellow at the base of the hind-wings. They are found in southern Queensland and in eastern New South Wales.

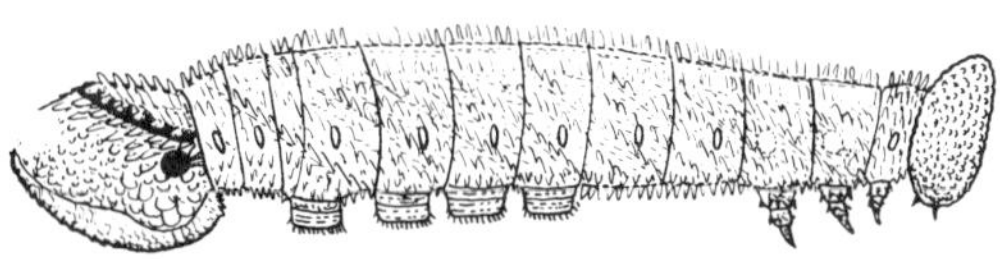

'Double-headed' larva of *Coequosa triangularis*. Length over 100 mm.

The humming-bird moths or bee hawks are found mostly in northern Australia and are both coloured and fly like bumble bees. Their forewings are often clear and free from scales. The caterpillars of these moths are sometimes found on gardenia bushes. The green and white striped privet moth caterpillar *(Psilogramma* species) is conspicuous for the large fleshy spine at the rear.

PROMINENTS (Family NOTODONTIDAE)

These are large moths with fat, furry bodies, a long forewing and a somewhat shorter hind-wing. Some have elaborate markings on the forewing and when at rest, the wings are held like a roof over the body. The caterpillars are large and often brightly coloured; when approached, they usually raise their tail defensively.

The banksia moth *(Danima banksiae)* typifies the group; adults have soft grey forewings with black and white markings and a span of some 75mm, the hind-wings are brownish-grey and the orange body is tipped with a brush of black hairs. Its caterpillars are large and fat and rich brown with black patches enclosing white spots. They feed on banksia bushes and trees. The caterpillars pupate underground.

Also included in this group are the Australian processionary caterpillars *Ochrogaster contraria* which live in silken nests on wattle trees. They emerge at night, in a procession to feed. The hairs of these caterpillars and related species can be intensely irritating if handled, or if a nest is opened by hand. When moving the caterpillars travel in a long line, each with its head in contact with the tail in front; as they travel, each caterpillar spins a thread of silk so that a narrow silken ribbon marks their passage across open spaces. Pupation occurs in a burrow in the soil.

The banksia moth. Wingspan 50 mm.

TUSSOCK MOTHS (Family LYMANTRIIDAE)

These fat and hairy moths have broad wings and many species are orange or white with various markings. The caterpillars are sometimes known as 'woolly bears' and are covered with dense hairs which can cause intense irritation if they contact human skin. Like the prominents (see p. 178) some of these caterpillars live in communal silken nests spun in the foliage of trees. Among the better known species, the painted apple moth *(Orgyia anartoides)* has 25mm span forewings, which are brown marbled with buff and grey; the hind-wings are bright yellow and black.

Tussock moth of *Iropoca* genus; male. Wingspan 40 mm.

Caterpillars of the mistletoe browntail moth *(Euproctis edwardsi)* whose adults reach 50mm wing-span, are found on mistletoe growing on eucalyptus trees. Their hairs are said to be exceedingly poisonous and irritating. At times, strong winds carrying fragments of these hairs have caused outbreaks of skin rashes in some areas of New South Wales.

TIGER MOTHS and **FOOTMEN** (Family ARCTIIDAE)

Often brightly coloured in shades of orange, red, black and white, these gaily coloured moths are night fliers. There are two subfamilies—the smaller and slender Lithosiinae or footmen and the larger, fat-bodied Arctiinae or tiger moths. The adult footmen have no ocelli, although they are present in tiger moths. Like tussock moth caterpillars, some caterpillars of this family are also covered with dense woolly hairs and may also be known as 'woolly bears'. Possibly the most common tiger moth is the species *Spilosoma curvata*, whose wings are marked in black, orange and red.

OWL MOTHS and **ARMY** or **CUTWORMS** (Family NOCTUIDAE)

With 1100 species known in Australia, this family includes a wide range of moths. Their name derives from the fact that night flying species have eyes which reflect light with an orange glow. Most are attracted by lights.

One of Australia's most famous species is the bogong or bugong moth *(Agrotis infusa)*, which is sometimes found in huge swarms in the Bogong high plains where they migrate and cluster in crevices in the rocks and in caves during summer months. These moths are dark brown or black and may swarm into houses and about street lights. When disturbed, they can release an unpleasant odour. The larvae of the bogong moth are known as 'cutworms' and vary from pale yellow to dark green with stripes. They infest tomatoes, maize and winter crops and pastures in south-eastern Australia and are a pest of considerable economic proportions.

Bogong moth. Wingspan 45 mm.

Owl moth, *Plusia chalcites*. Wingspan 35 mm.

DAY MOTHS (Family AGARISTIDAE)

The largest species, the painted vine moth *(Agarista agricola)* has black front wings marked with red, yellow and blue, and is found on the east coast, north of Sydney. One of the most unusual moths is the Australian whistling moth, *Hecatesia fenestrata*. The 'whistling' sound is made only by the males. When flying, a transparent ribbed area on the forewing is rubbed against a small protrusion, and this makes a low whistling sound. If disturbed the moth stops whistling and flies away quite silently. Whistling moths are found on the coast and tablelands of eastern Australia; similar species are found in the south-east and south-west of the continent. The whistling sound may often be heard at dusk in these regions. Some of the well-known vine moths are serious pests to viticulture.

Day moth.

Butterflies

To date 359 species of butterfly have been described in Australia. These are grouped into six families which may be broadly identified by the following features:

Head	*Front legs*	*Wings*	*Antennae*	*Flight*	*Caterpillars*
Wider than thorax	As long as others		Set wide apart at base with hooked clubs at tip	Short and darting	Tapered at both ends . . .**Skippers**
Narrower than thorax	As long as others	Have tail projections, generally large wing-span		High above ground	Fleshy and smooth, generally have forked scent organ . . .**Swallowtails**
	As long as others	Predominantly white or yellow			Long, covered with short white hairs but without fleshy spikes or tubercles . . .**Whites** and **yellows**
	Slightly shorter than others on males	Often have tails; frequently blue, usually small	Have long clubs		Look like hairy slugs, head may retract into body. Often attended by ants. . .**Blues, coppers** and **hairstreaks**
	Shorter than others		Labial palpae form distinct beak		Small heads and covered with fine hair; found only in N. Australia. . .**Beaks**
	Poorly developed and hardly noticeable				Bear fleshy spines or tubercles. . .**Satyrs** and **nymphs**

DARTERS and **SKIPPERS** (Family HESPERIIDAE)

With a distinctive darting form of flight which gains these butterflies their name, the skippers are primitive butterflies in some ways similar to moths, both in their wing venation and in their dull colouring. Most skippers fly by day although some also fly at dawn and dusk. One of the most primitive skippers is the regent skipper *(Euschemon rafflesia)*.

The male of this butterfly has a frenulum or spine which fastens the forewing and hind-wing together in flight. Females of the species have no such frenulum.

Skippers may also be recognised by their large heads (wider than the thorax) and their large protruding eyes. The antennae are set wide apart at the base and usually have a hooked club at the tip. The body is thick and all six legs are well developed. At rest, the wings are usually held above the body but some species rest with them in a horizontal position.

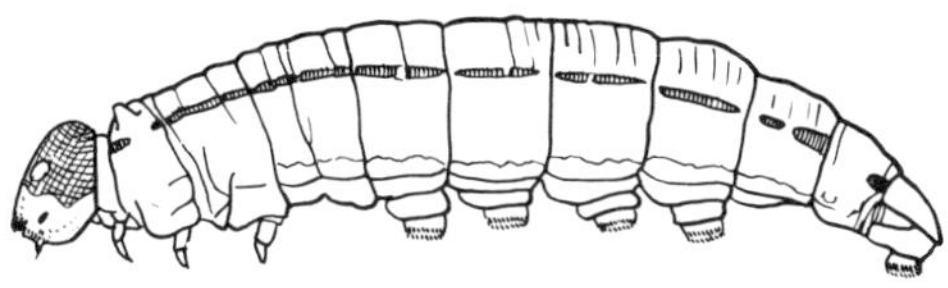

Larva of *Euschemon rafflesia*. Length 38 mm.

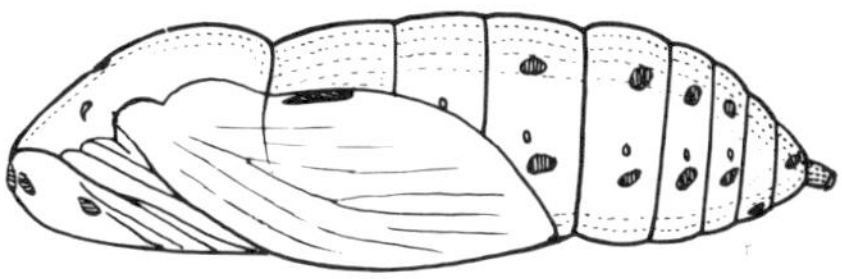

Pupa of *Euschemon rafflesia*. Length 31 mm.

Caterpillars are long and hairless, and usually live in a shelter formed by a curved or folded leaf or by two leaves joined with silk; here pupation takes place. Both larvae and pupae are covered with white powder. Caterpillars have hard heads and often distinctive markings.

Some 100 species of skippers have been described in Australia, most inhabiting the warmer east coast regions, but extending into Victoria. Those which rest with wings laid flat like moths are known as 'flats' and another group are known as 'awls' as the third joint of the sense organs near the mouth *(palpi)* is shaped like an awl. Species in the genus *Chaetocneme* often have bright red eyes and are found in eastern Queensland and the northern half of New South Wales.

Caterpillars of the orange palmdart *(Cephrenes augiades)* are pests of ornamental palms in eastern Australia. In the wild, they are found on bangalow and cabbage-tree palms, and the caterpillars rest between two fronds bound together with silk. Many species feed on sword grass.

Typical skipper.

SWALLOWTAILS (Family PAPILIONIDAE)

Most species of swallowtails have a tail-like projection on the hind-wings and are often large. Although they often fly high above the ground, they may be captured when feeding on flowers. Most are found in the warmer parts of Australia. The fleshy swallowtail caterpillars are large, often humped towards the front and may have a series of protuberances in their younger stages. When disturbed, the caterpillar can project a forked fleshy scent gland *(osmeterium)* from a slit in the thorax which also emits a pungent odour.

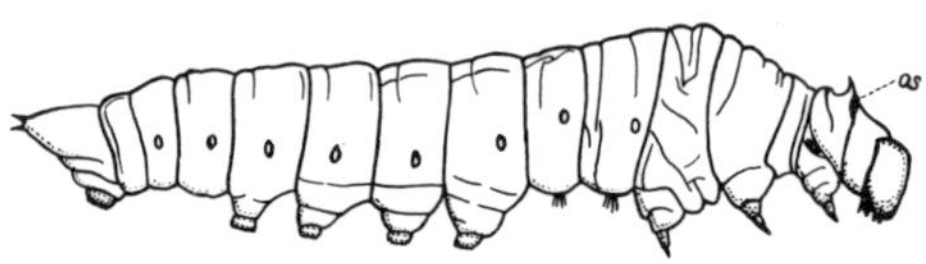

Swallowtail caterpillar, showing opening of *osmeterium*.

Of Australia's 18 species of swallowtails, the Cairns birdwing *(Ornithoptera priamus)* is the largest butterfly in Australia, with the female attaining a wing-span of 190mm.

Caterpillars of the big greasy *(Cressida cressida)* feed on poisonous Dutchman's pipe vines, while those of the orchard butterfly *(Papilio aegeus)* are a pest of citrus trees on which they feed.

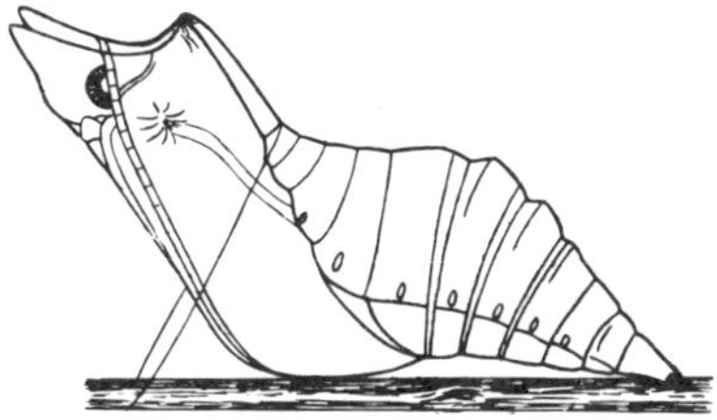

Pupa of *Papilio aegeus* showing silk girdle attachment. Length 45 mm.

Perhaps the most conspicuous swallowtail is the Ulysses butterfly *(Papilio ulysses)* which is found in north-eastern Queensland. It is black and bright metallic blue and has distinctive tails to the hind-wings. The Ulysses may grow to a wing-span of 110mm.

Male Ulysses butterfly. Wingspan over 100 mm.

WHITES and **YELLOWS** (Family PIERIDAE)

With a wing-span averaging 36 to 75mm, most of these butterflies are predominantly white or yellow. Distinguishing marks are often present on the wings of males. Many butterflies in this family have bright markings on the underside of their wings.

Among the 30 species described in Australia, yellows are grouped in the Coliadinae subfamily and whites in the Pierinae. Caterpillars are long, often green and usually covered with short white hairs. Some species spin a communal web of silk.

One of the most common white species is the imperial white butterfly *(Delias harpalyce)* with a wing-span of 75mm; the caterpillars live in mistletoe. Larvae of this species spin a communal web over twigs and leaves of the host plant and the pupae link themselves to the web. Adults are an off-white above and have orange and scarlet markings under the wings.

The migrating caper white butterfly *(Anaphaeis java)* lays conical eggs in wild orange trees in inland areas. Caterpillars are a dark orange green and may defoliate the bushes. Adults emerge in large numbers and then fly in flocks towards the coast. They are often seen in the Sydney and Melbourne regions from September to early December.

The canary-coloured grass yellow butterfly *(Eurema hecabe)* is distributed widely and is one of the first butterflies to appear in springtime in inland areas.

Possibly the best known white is the introduced cabbage white butterfly *(Pieris rapae)* found throughout eastern Australia. Its caterpillars can cause considerable damage to garden plants and vegetables unless preventive measures are taken quickly where evidence of their presence is found.

NYMPHS and others (Family NYMPHALIDAE)

Possibly the best identification features of this group of butterflies are the long hind-legs, and the short hairy forelegs. Seven subfamilies encompass 80 species described from Australia.

The group has a wide diversity of colour, shape and size, although the eggs are usually ribbed and strongly sculptured and domed. Caterpillars are often spiny, or have long paired horns or other slender growths.

The wanderers (subfamily Danainae) are famous for their migratory behaviour in North America. They are also found in Australia and similarly overwinter in warmer places in large flocks, resting on trees. The North American wanderer *(Danaus plexippus)* may be found anywhere along the eastern and southern coasts from Cairns to Melbourne and Adelaide. Caterpillars feed mainly on milkweed. All wanderers possess distasteful properties and their bright colours act as a warning to potential predators. They are also surprisingly resistant to injury and can withstand damage which would kill many other butterflies.

Typical Danainae caterpillar. Length 50 mm.

Wanderer butterfly.

The crows (genus *Euploea*) are mainly tropical and may be bluish-black, or dark brown with white markings. The oleander butterfly *(E. core corinna)* has silver or gold patterned caterpillars which, as its name suggests, feed on oleanders and fig trees.

The browns (subfamily Satyrinae) are usually orange and black with eye spots, and are weak fliers. The caterpillars feed on grasses and herbs.

The swordgrass brown butterfly *(Tisiphone abeona)* grows to a wing-span of 60mm and is found in coastal Queensland and New South Wales. The caterpillars are pale green and are usually found on swordgrass. The evening brown *(Melanitis leda)*, may attain a wing-span of 90mm and is found in northern and eastern Australia. Its caterpillars are green with two hairy horns and are pests of sugar cane.

Tailed emperor butterflies (subfamily Charaxinae), are large, fast flying insects with two pointed tails on each hind-wing. The caterpillars are green with yellow markings on the side and their heads have four horns. They feed on tree foliage, including various types of wattles.

Top: The swordgrass brown butterfly. Wingspan 55 mm.

Bottom: Tailed emperor butterfly. Wingspan 70 mm.

The owls (subfamily Morphinae), are found only in Cape York Peninsula and the Torres Strait islands. These are shade-loving butterflies whose name derives from the two large eye spots beneath the hind-wings. Caterpillars have two horns on the head and a forked tail. They feed on palms.

In the nymphs (subfamily Nymphalinae) possibly the best known species is the Australian painted lady *(Vanessa kershawi)*, a brown and orange butterfly with four rather small blue eyes in the orange part of the hind-wings. The forewing is black with orange markings and white spots near the tip. Its caterpillars are brown or black and are hairy. They are often found on everlasting daisies and cape weed. The pupa is brown, sometimes showing silver or gold spots.

Another common nymph is the meadow argus *(Precis villida)* which is found in many parts of Australia and is mainly brown with a broad wavy orange band. There is a large eye spot in the centre of the orange on the forewing and there are two smaller eye spots on each hind-wing. Caterpillars are black with short spines.

Glasswings *(Acraea andromacha)* are common in Sydney and further north and have almost transparent forewings marked only by a few patches of brown scales. The hind-wing is white with some black spots. The caterpillars feed on wild passionfruit vines and are brown with long spines along the back.

The 'glasswing' butterfly.

BEAKS (Family LIBYTHEIDAE)

Closely related to the nymphs, only one Australian species of beak is known. This is the Australian beak *(Libythea geoffroyi)*, present in the northernmost regions of the continent. They are best known for their habit of resting under the eaves or verandahs of houses during the heat of the day. Their name comes from the long, beak-like tongues *(labial palpi)*. Males have short forelegs, like the nymphs, but in the females they are functional. Caterpillars have small heads and are covered with fine hairs.

COPPERS, BLUES and **HAIRSTREAKS** (Family LYCAENIDAE)

Butterflies in this family often have eyes notched at the base of the antennae. The forelegs of males are slightly shorter than the others but are normal in females. On the hind-wings, there are often one or two hair-like tails. These butterflies are active in the day and are found mostly in the tropics. Caterpillars often secrete certain fluids in a honey gland which ants collect as food; some species even live in ants' nests.

Many of these butterflies are brilliant blue and orange (or dull brown). They have a characteristic quick, short flight.

Typical hairstreak butterfly.

In the subfamily Liphyrinae, the moth butterfly *(Liphyra brassolis)* is black blotched with orange in its adult form. It is best known for its caterpillars, which are very flat and hardened and live in the nests of the aggressive green tree ants *(Oecophylla smaragdina)* where they prey on the ant larvae, sucking out their body juices. Because these caterpillars are so flat, the head and legs cannot be seen unless they are turned over. The caterpillars have a very hard outer surface which cannot be pierced even by the jaws of adult ants. Pupation occurs in the ants' nests and when butterflies emerge, they are protected from attack by the ants by hard white scales on the wings and thick fur on the abdomen. The wing scales are lost when the butterfly is ready for flight.

Blues (subfamily Lycaeninae) usually have blue wings, though in some species they are orange. Most species have honey glands and may be associated with green tree ants or with meat ants.

Blue butterfly of *Thysonotis hymetus* species, male. Wingspan 30 mm.

Female of same species.

Found from south Queensland to Victoria the common imperial blue *(Jalmenus evagoras)* lays its eggs on wattle trees. These are found with black ants. When pupation is due, the caterpillars spin a communal web among the twigs and the shiny black pupae are housed within this.

The fiery jewel butterflies (*Hypochrysops* species) are found in eastern Queensland. They have bright orange or purple wings with metallic colours underneath. Those species whose caterpillars eat acacia leaves or mangroves have their eggs protected by ants which cover them with debris. When caterpillars hatch, the ants feed them and when large enough to forage for themselves, the ants lead the caterpillars to the most tender shoots. After feeding, the ants herd their caterpillars back to the 'byre' usually at the base of the plant. Should a caterpillar stray on its journey, the ants will push the caterpillar's head in the direction it should travel. These ants also protect the caterpillars from enemies. In return for this careful tending, the ants are able to stroke glands of the caterpillars and obtain a sweet honey. Eventually, pupation takes place in the byre and when the butterflies emerge, ants may often be seen crawling all over them without apparently causing any discomfort.

According to McKeown, the common, bean or lucerne blue *(Zizeeria otis)* is possibly the most common butterfly in Australia. With a wing-span of only 25mm, they may be seen in their thousands fluttering through lucerne paddocks. Caterpillars feed on lucerne, clover and the buds and pods of French beans.

Note: For more detailed guides to the various species, readers are referred to *Australian Butterflies* by I. F. B. Common (Jacaranda Pocket Guide, 1964); *Butterflies of Australia and New Guinea* by Charles Barrett and A. N. Burns (Seward, 1951); *Australian Butterflies in Colour* by Alexander Burns (Reed, 1969) and *Butterflies of the Australian Region* by Bernard D'Abrera (Lansdowne, 1971).

SAWFLIES, WASPS, BEES AND ANTS
Order HYMENOPTERA (membrane-winged)

In some ways likely to be confused with flies (Diptera) because of their stiff, membraneous and transparent wings with a reduced number of veins, Hymenoptera differ from Diptera in having four wings instead of two, although the hind-wings may often be much smaller than the forewings. In flight, the wings are usually joined by minute hooks *(hamuli)* on the front edge of the hind-wings, which engage in a fold on the trailing edge of the forewings.

With the exception of sawflies (Symphyta), Hymenoptera have a narrow waist separating the thorax from the second, and subsequent segments of the abdomen *(gaster)*. Females all possess ovipositors which may be adapted for sawing (sawflies), piercing (ichneumons) or defensive stinging (wasps, bees and ants). Males possess no ovipositor and cannot sting. Most Hymenoptera have chewing mouthparts though some may also be adapted for sucking. Metamorphosis is complete (egg, larva, pupa and adult) though some larvae pass through several distinct stages.

In Australia there are nearly 9000 designated species of Hymenoptera. These are most easily considered grouped by their various common names and the following section will deal with sawflies (suborder Symphyta), ichneumons (superfamily Ichneumonoidea), wasps (which are included in several superfamilies), bees (superfamily Apoidea), and ants (superfamily Formicoidea).

As a group, Hymenoptera are possibly the insects of greatest benefit to mankind. While some are nuisances or pests (such as stinging wasps and ants which enter houses or food stores) the number of species which benefit man far exceeds those which are pestilential.

Among the wasps, many are predators on a wide variety of insects and other arthropods, including many pests. Bees are invaluable in pollinating thousands of varieties of plants, fruit trees and other crops and are valuable as honey producers. Ants aerate the soil and are also useful scavengers.

Some Hymenoptera have highly organised social structures, which may be likened to that of termites (see pp. 13–20) although others lead solitary lives.

For identification, the sawflies (suborder Symphyta) may be separated from other Hymenoptera (suborder Apocrita) by the following features:

	Sawflies	*Other Hymenoptera*
Base of abdomen	Not constricted (without a 'waist')	Constricted as base between first and second abdominal segments ('waisted')
Ovipositor	Adapted into a saw-edged apparatus for cutting into plants	Not adapted into a saw edge
Larvae	Caterpillar-like but without abdominal legs or, if present, lacking circle of hooks on abdominal legs. Only one ocellus on each side of larva's head (caterpillars have several)	Maggot type of grub, always without legs.

There is some confusion for the layman in the modern scientific arrangement of wasps. Sawflies are distinctive for lacking a waist, but nevertheless have wasp-like behaviour. Similarly, ichneumons, while now grouped with wasps, have long been considered a separate insect type. Furthermore the various types of 'true' wasps have now been grouped in orders based on modern scientific classifications. While these are scientifically valid, the older method of identifying these insects is possibly easier than the more technical, modern methods.

For this reason, the section on 'wasps' will include sawflies, ichneumons, and wasps of various kinds. Like ichneumons, many of these are parasitical on other insect larvae and perform a valuable service to mankind. At present, more than 6000 species are classified as sawflies, ichneumons and wasps in Australia and they include a large variety of insects. Here, only the more common families are described.

SAWFLIES (Suborder SYMPHYTA)

Lacking a conspicuous waist, female sawflies have a delicate saw-like ovipositor which is housed in a slot at the end of the abdomen. This 'saw' is used to cut a slit in the leaf or the stem of a plant, into which eggs are deposited. The eggs then hatch into larvae which cluster together and feed in a mass.

Female sawfly.

Clusters of steel-blue sawflies (*Perga dorsalis* and *P. affinis*) are known as 'spitfires'. If disturbed, they wave the abdomen and exude a strong-smelling liquid from their mouths. The larvae eat the soft tissues of various leaves and can cause much damage to eucalypts and other plants. When larvae are mature, they go to ground and burrow in the earth where they pupate in a tough silken cocoon.

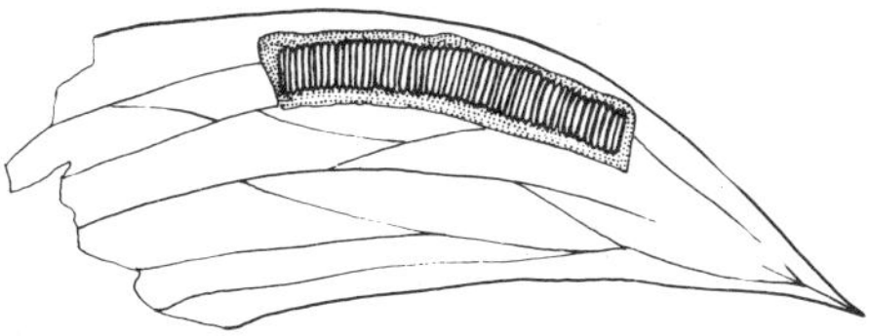

Eggs of *Perga dorsalis* in leaf.

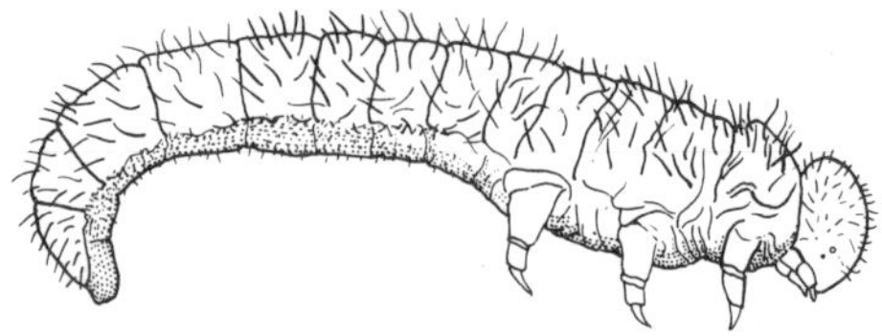

Larva of *Perga dorsalis.*

The female of the genus *Pergagrapta*, a slender reddish-yellow insect, mounts guard over her eggs and protects them like a broody hen, spreading her wings, opening her jaws and buzzing loudly if an intruder approaches. Females of this species are easily caught.

Larvae of the introduced species *Caliroa cerasi* are known as the pear and cherry slugs. The larvae are soft-bodied and slimy and are pests of some fruit trees and shrubs. They defoliate the plants and cause a 'burnt out' look in fruit trees.

SIREX WASPS (Superfamily SIRICOIDEA: SYMPHYTA)

Related to the sawflies, and now a serious introduced pest of softwoods (particularly *Pinus radiata*) sirex wasps *(Sirex noctilio)* are large insects with metallic colours and an abdomen often ending in an upturned spine or horn. The females deposit eggs and fungi spores through the bark. The larvae tunnel into the wood. At the same time, the rot-producing fungus enables them to eat the wood. This fungus eventually kills the tree, the larvae and the fungus being mutually dependent.

ICHNEUMONS (Superfamily ICHNEUMONOIDEA)

Sometimes known incorrectly as ichneumon 'flies' or 'wasps', ichneumons are parasites, depositing eggs in the immature stages of many different types of insects. Adults usually have long legs, an abdomen *(gaster)* which is curled up from the thorax and a very narrow waist; they are often banded. Ichneumons have long and slender antennae and a particularly distinctive wing venation. They are generally between 30 and 40mm in length and to date some 400 species have been described in Australia.

One of the most common species is *Netelia productus* which is attracted to light and may often enter a house on a warm summer evening. It is a bright orange-yellow, with its gaster flattened at the sides. It has three large simple eyes *(ocelli)* and long legs. This species mostly parasitises caterpillars (Lepidoptera) and cutworms but may also attack sawfly larvae. Females can inflict a perceptible but not very painful sting on man. The spotted ichneumon *(Echthromorpha intricatoria)* is blue-black with a double row of white spots along the back.

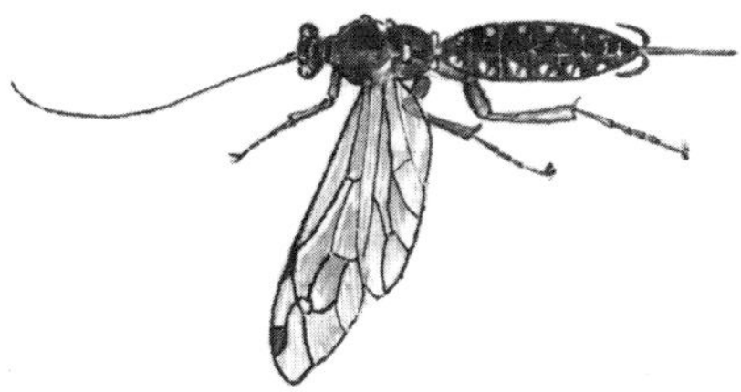

Echthromorpha intricatoria, a common Ichneumon. Wingspan 45 mm.

GALL-FORMING WASPS (Superfamilies CYNIPOIDEA, CHALCIDOIDEA)

Generally small (less than 6mm long) although some may attain 25mm, gall-forming wasps deposit eggs in plant leaves and stems. These cause swellings or galls similar to others caused by Hemiptera. Various native and introduced species form galls on eucalypts, acacias and European oak trees. Identification is based on the number of antennal segments, wing venation and other features. The ovipositor is usually hidden within the gaster.

In the superfamily Chalcidoidea, the fig wasps (Agaonidae) include the Moreton Bay fig wasp *(Pleistodontes froggatti)* which has a strangely lengthened head. This insect is solely responsible for fertilising this fig tree. Wasps in the genus

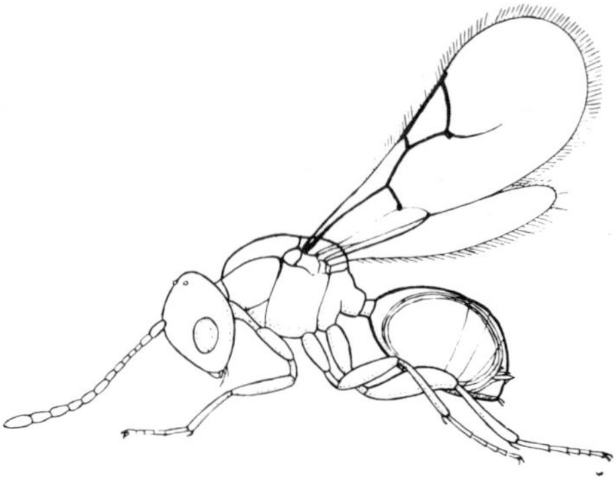

Gall wasp, showing ovipositor coiled within gaster (end section of abdomen).

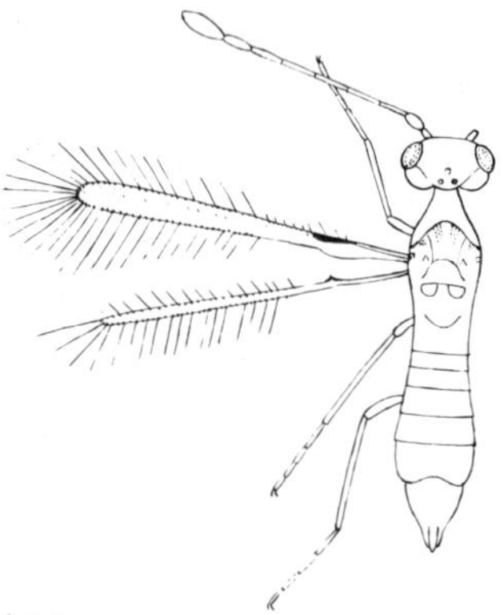

Fringe-winged *Mymar* genus wasp.

Mymar have very slender, fringed wings. Others have hind-legs adapted to jumping.

Other species may cause gall formation in lucerne (introduced *Bruchophagus gibbus*), citrus trees (*B. fellis*) and other plants. Overseas, the chalcid *Alaptus magnanimus*, a parasite on booklice, never exceeds a length of ·2mm and is the smallest recorded insect in the world.

EMERALD or **CUCKOO WASPS** (Family CHRYSIDIDAE)

Usually a rich metallic green or blue, these wasps may also be purple or reddish. The body is covered with a thick casing. The adults lay eggs beside the larvae of mason and sphecid wasps. To protect themselves against the returning host, the adult emerald wasp can roll itself into a protective ball. The largest and most beautifully coloured Australian species is *Stilbum splendidum*, a blue-green insect with an almost unnoticeable waist.

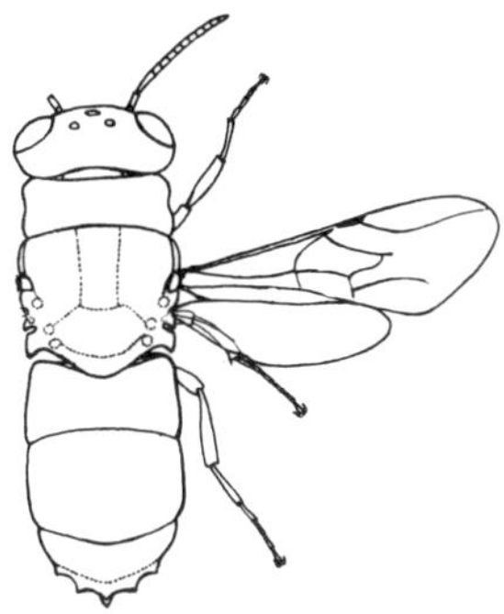

Emerald wasp.

SPIDER HUNTERS, SAND or **GROUND WASPS** (Family POMPILIDAE)

Active, smooth-bodied insects with long legs, these wasps have wings which are tinged with yellow or marked with black or dark brown. They dig burrows in sand scratching it away with the forelegs and throwing it back between the hind-legs. To stock the burrow with food some of these wasps hunt spiders which they sting into immobility and carry back into the burrow. They then lay an egg into the paralysed prey. Possibly one of the best known species is the large orange and black wasp in the genus *Cryptocheilus* which may grow to a length of 35mm.

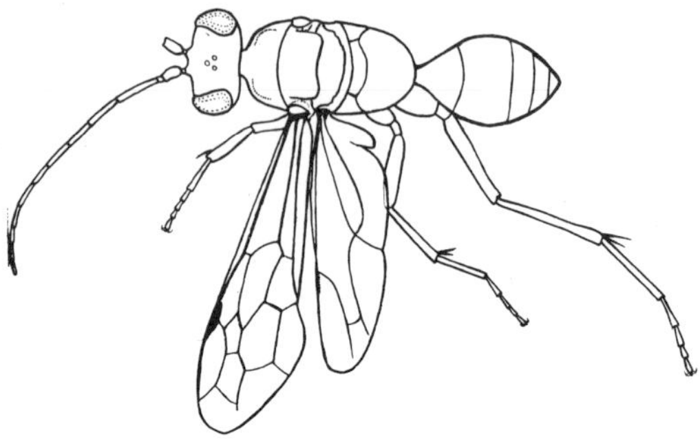

Ground wasp.

Species in the genus *Phanagenia* build mud cells under tree bark while *Aulacostethus* paralyses the spider in its nest and deposits the egg in it there.

FLOWER WASPS (Families SCOLIIDAE, TIPHIIDAE, MULTILLIDAE)

The Scoliidae are large, hairy insects and many have rich-blue bodies and wings. They are also parasitic wasps, affecting the larvae of scarab beetles. One of the largest species is *Trisciloa ferruginea* which has a rich-golden downy covering.

In the family Tiphiidae most females are wingless with short curled antennae. They mate in mid-air, the male carrying the female and then dropping her to the ground. There the female of *Diamma bicolor* deposits her egg in an immature mole cricket. She is often wrongly called a 'blue ant'. Slender-winged male tiphiid wasps have large compound eyes and often alight on flowers, or may be found near flowering trees and shrubs.

Females of the Mutillidae are also wingless and their hairy bodies have earned them the name 'velvet ants'. They parasitise burrowing wasp and bee larvae. Most are grouped in the genus *Ephutomorpha* and their length ranges from 5 to 25mm.

Male 'velvet ant', *Ephutomorpha ferrunginata*. Wingspan 30 mm.

Female of same species.

SOCIAL or **PAPER-NEST WASPS** (Family VESPIDAE)

Polistes wasps are known for their mushroom shaped paper nests hung from the dome by a slender stalk and often found in the eaves of buildings. The single layer of hexagonal cells are arranged in a regular order on the under-surface, rather like a honeycomb. The best known species, *Polistes variabilis* and *P. humilis*, have a painful sting. The nest is made from wood fibres mixed with saliva and may measure up to 230mm in diameter.

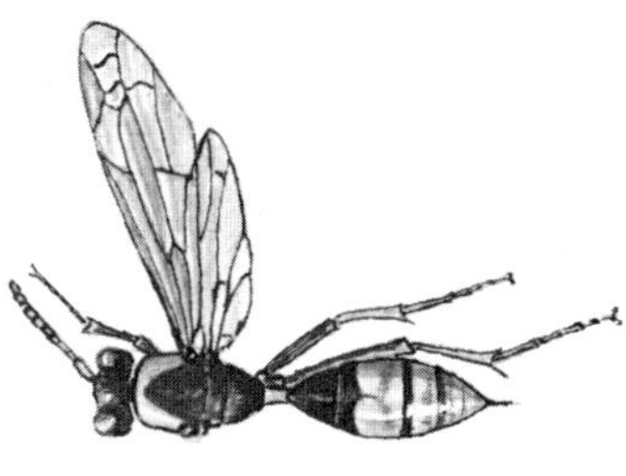

Paper-nest wasp of *Polistes* genus.

Some paper wasps have social castes and workers feed the larvae with masticated caterpillars. When the larvae are ready to pupate, they spin a silken screen over the cell entrance. At rest, adult vespids fold their forewings lengthways. In the genus *Ropalidia*, wasps build nests which have many layers and are often enclosed in an external casing of paper. The North Queensland species *Ropalidia cabeti* is a small yellowish species which builds nests on the lee side of a branch, thus protecting it from rain. The introduced English wasp *(Vespula germanica)* makes a similar nest.

MUD-DAUBERS, MASON or **SOLITARY WASPS**
(Family EUMENIDAE)

Carrying their wings folded along the top of the body when at rest, the female mason wasps build nests of clay, often on the walls of buildings. In each cell of the nest the female places a paralysed caterpillar. In this she lays her egg. The cell is then sealed with a plug of clay and other cells are built alongside and beyond the first cells. When the egg deposition is complete, the female sheaths the nest with a thick coating of clay, completely sealing her own entrance tunnel. Wasps of the genus *Abispa* build mud cells inside homes in Queensland. This wasp is large with an orange-yellow body with a black band around the gaster. Wasps in the genus *Eumenes* grow to a size of 30mm and have a long, slender waist. They make mud nests in inland locations.

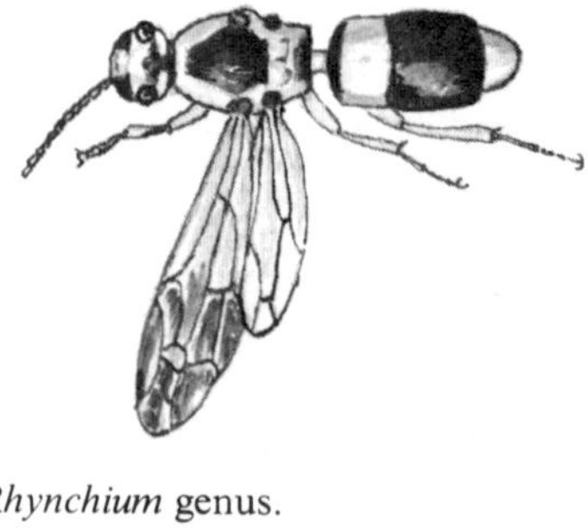

Mud-dauber of *Rhynchium* genus.

CLUB WASPS or **CICADA HUNTERS** (Family SPHECIDAE)

With exceptionally long and strong legs, the cicada-killing wasp *(Exeirus lateritius)* is yellow and black with orange-tinted wings. It digs large burrows in sandy soil and the mouse-hole size entrance is usually surrounded by a mound of soil. Such burrows extend about a metre underground and branches off the main tunnel lead to individual cells.

Many wasps in this family prey on specific insects and the size of the prey depends on the size of the particular species of wasp. To date, 420 species of these predatory sphecid wasps have been described in Australia; their prey includes spiders, caterpillars, grasshoppers, beetles and many other insects.

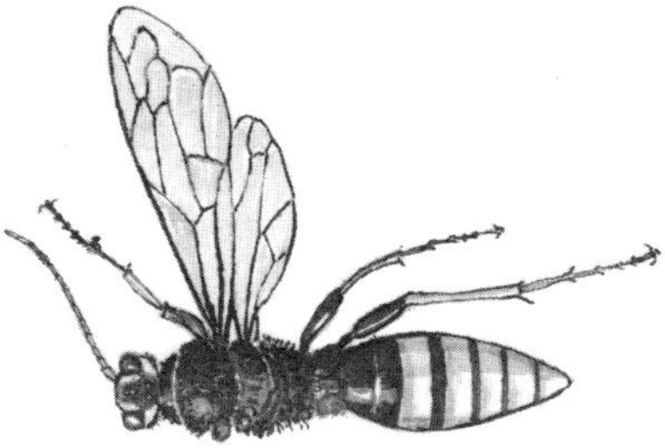

The cicada-killing wasp, *Exeirus lateritius*.

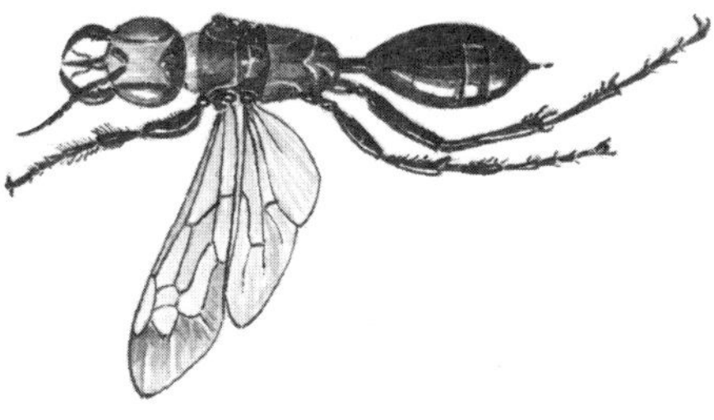

Predatory sphecid wasp.

BEES (Superfamily APOIDEA)

It has been estimated that some 3000 species of bees occur in Australia, although of these only 1600 have so far been described. The world species population is estimated at 20 000 with sizes ranging from 2mm to 39mm in length.

While the well-known domesticated bees live communal lives and have highly-organised social activities, the majority of native bees (including the Australian species) are solitary insects. Each female makes her own nest and there are no separate queen or worker castes. Nests are burrows in earth, plant stems or wood, or sometimes are made from resin or mud in exposed locations. The female makes the nest and lays the eggs, but usually dies before the young emerge. Unlike bees in other areas, most Australian native bees are less than 12mm long and are dependent on a particular family of plants from which they gather nectar.

Bees may be recognised by their hairy legs, although the location of pollen brushes *(scopa)*, or pollen baskets *(corbicula)* consisting of a fringe of hairs around a hollow, are the most important features of identification.

Some bees' tongues are long in order to 'sip' nectar. Scientists consider that bees are in fact wasps which have adapted from an animal diet to feed on plant products. Larvae are fed with a combination of pollen and honey. By transferring pollen from plant to plant, bees are valuable for their pollination of many types of plants.

Identification of bees

When present, pollen brush at end of abdomen
. . . **Leaf-cutters** (MEGACHILIDAE)

When present, pollen basket on tibia of hind-legs
. . . **Social bees** (APIDAE)

Pollen brush present on hind-legs
. . . **Carpenter bees** (ANTHOPHORIDAE)

Fine hairs on crescent-shaped hind tibia *Ctenoplectra* sp. of N. Queensland . . . (MELITTIDAE)

Solitary bees digging burrows lined with transparent material . . . **Burrowing bees** (COLLETIDAE)

Pollen brush in various positions, burrows in ground, sometimes communal, some species parasitic . . . (HALICTIDAE)

LEAF-CUTTING BEES (Family MEGACHILIDAE)

Ranging from very small to 25mm in length, these bees are noted for their large square heads and cylindrical bodies. There is often a large tuft of hair at the tip of the gaster or abdomen. Nest-building species build a complicated nest, sometimes in a burrow, or on a tree trunk, fence post, or wall. The nest is lined with or made from regularly cut fragments of rose leaves (genus *Megachile*) or resin (genus *Chalicodoma*). Burrows and cell partitions of the genus *Lithurge* are made of clay. Larvae of these long-tongued bees all spin strong cocoons before pupating.

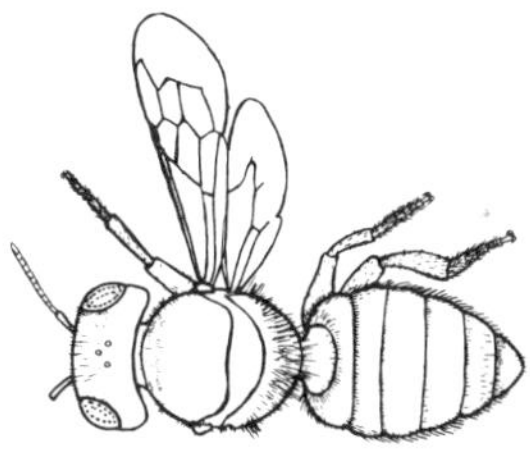

Leaf-cutting bee.

SOCIAL BEES (Family APIDAE)

Best known for the hairy and introduced honey bee *(Apis mellifera)* this family contains bees which have a strong social order. The Australian native social bees (genus *Trigona*) include the small dark native bees which are stingless and form colonies in cavities of trees. The watery honey from these nests is a valuable part of the diet of some aborigines and was once the only reliable source of sugar for outback settlers.

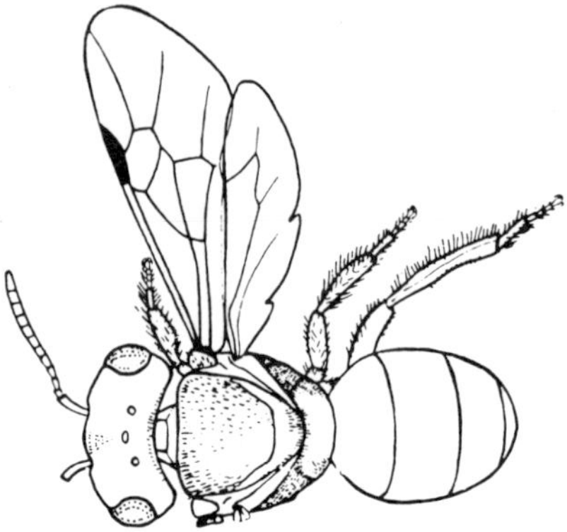

Australian native social bee, *Trigona carbonaria.*

CARPENTER BEES (Family ANTHOPHORIDAE)

Some of the bees in this family are small and have few hairs while others may be quite large and hairy. Some species are parasitic on other bees. Probably the best known are carpenter bees (subfamily Xylocopinae) which excavate their nesting chambers in the stalks of grass trees *(Xanthorrhoea)* and in dead timber and reeds. These large hairy bees are metallic blue, black or green. The males have a large tuft of golden hair on the thorax and resemble the bumble bee of overseas. When visiting flowers, carpenter bees also buzz like bumble bees. The nest is excavated by the bee cutting an entrance hole into the stem and then burrowing downwards. Using fragments of pith and saliva, the bee separates its tunnel into cells and in each is left a ball of pollen and an egg. When the larva hatches, it lives on this 'bee bread' and eventually

Carpenter bee.

pupates without a cocoon. Australian species in the genus *Nomada* are found in Queensland and are somewhat wasp-like and parasitic. Banded bees (genus *Amegilla*) nest in large numbers in intersecting burrows in earthen banks. Cells are lined with wax.

SOLITARY BURROWING BEES (Family COLLETIDAE)

These bees vary in size and in hairiness; all dig burrows which they line with a sort of acetate. Many bees have pollen brushes but carry pollen in the mouth or crop. While bees in the subfamily Colletinae nest in the ground, those in the Hylaeinae burrow in pithy stems, or in dead logs and stumps. All Australian bees in this family are solitary. The most common species are *Paracolletes tuberculatus* and *P. obscurus*.

Solitary burrowing bee.

SOCIAL BURROWING BEES (Family HALICTIDAE)

Medium to small, with black or green bodies (sometimes with red markings) these bees burrow both in soil and in rotting wood. Often several fertile females may nest in a single burrow; some species have workers.

ANTS (Family FORMICOIDEA)

Australia is now known to have at least 1100 species of ants, which is about 10 per cent of the world total. These species are grouped into eight subfamilies and include the Australian bulldog ants *(Myrmecia)* which are considered to be among the most primitive of all ants and are restricted to the Australian region. They are also the largest ants in the world; the Queensland bulldog being up to 35mm in length. Ants are social insects, and form male and female castes. Females, in turn, are divided into fertile females, which found the nest or colony, and infertile females which may become major and minor workers or soldiers. It has been suggested that an ant develops the characteristics of a particular caste according to the quantity and composition of food given to larvae. Fertile queens begin life with wings, but workers and soldiers are always wingless. Soldiers have large heads, workers usually have strong jaws and may be armed with stings, or have the capacity to squirt formic acid into wounds caused by their jaws. Males have permanent wings and relatively small heads with large compound and simple eyes. Their jaws are simple and often seem to have no function. In a colony, the queen is the largest insect, the males next in size, followed by the soldiers and workers.

When the time comes to establish a new colony, winged males and females swarm, mate in flight, and then fall to the ground. Males die and females lose their wings and seek shelter, often digging a small tunnel. There the first eggs are laid and the larvae are fed by the queen; but after pupation the young become workers, seek food and feed the queen, and excavate the chambers and passages of the nest either in the ground, or in timber; some species build their nests among foliage. While some nests have inconspicuous entrances, those of some bulldog ants, and the mulga ant *(Polyrhachis macropus)* have distinctive mounds or craters. The nests of mound, or meat ants *(Iridomyrmex purpureus)* have gravel caps extending up to 3m in diameter; from them tracks beaten by hundreds of ants extend in various directions and often for a long way. These tracks usually lead to food sources. When disturbed, meat ants will bite fiercely, often releasing a repellent smell.

Although many ants are particularly aggressive, some will permit *inquilines* or intruders to live in their nests unharmed. These may include the caterpillar of the butterfly

Liphyra brassolis (see pp. 190–1) which lives in the nest of the green tree ant *(Oecophylla smaragdina)*. Some 200 different types of beetles, flies, cockroaches, silverfish and other insects have been found as inquilines in nests of various ants.

Many ants are useful to man, acting as scavengers, soil aerators and as predators of insect pests. However, some species are themselves annoying pests. These include the introduced Argentine ant, meat ants which may infest food processing plants, and those species which inflict severe stings.

Ants may be broadly divided into two groups. The first is identified by having two knobs on the 'waist' *(gaster)* joining the abdomen to the thorax. The others have only one knob or *node*. Other features include a conspicuous sting, visible on dead specimens, or its absence. Only four of the subfamilies are described in detail here.

BULLDOGS and **JUMPERS** (Subfamily MYRMECIINAE)

With 65 species described to date, bulldog ants (genus *Myrmecia*) or bull-ants are conspicuous by their large size. The best known species are the red and black bull-ants *(Myrmecia gulosa* and *M. negrascapa)* which have a red head and body, a black apex to the abdomen and slender yellow jaws. A queen may measure up to 25mm in length. There is also the glossy black *M. tarsata* which has a red tip on its abdomen. Jumpers or 'jumping jacks' build mound nests and will attack anything that comes near it, often jumping 150 or 200mm. Once angered, they are hard to stop. The most common jumper is black and yellow *(M. pulosa)* about 13mm long and more slender than a bulldog. Jumpers are more aggressive than the bull-ants and possess abdominal stings which can inflict a painful wound. On-the-spot relief is said to come from rubbing a sliced root of bracken or fern on the bite.

Bull-ants have underground nests and pupae are enclosed in cocoons of papery brown silk. Their elongated oval shape sometimes causes them to be named 'ants' eggs', but the

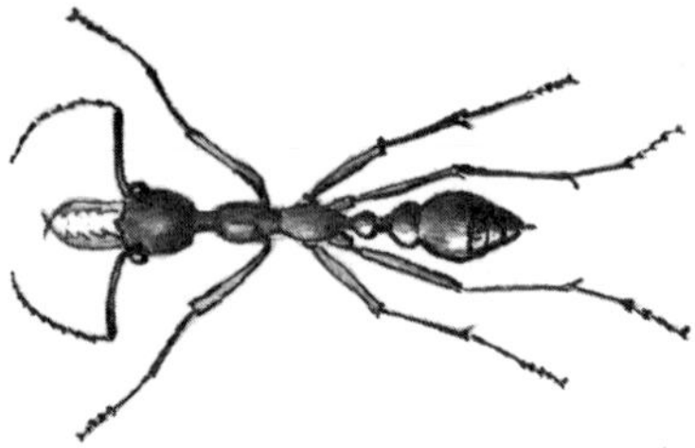

Myrmecia gulosa, worker. Length 25 mm.

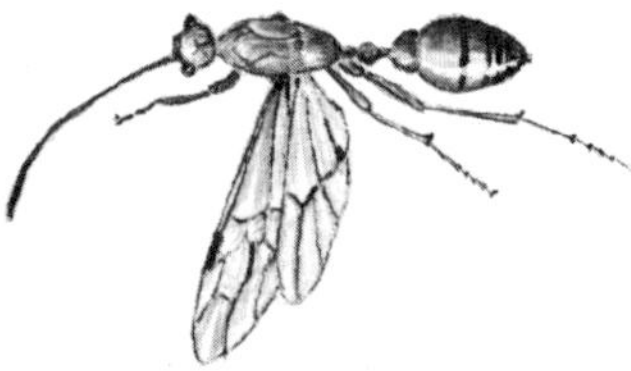

Male, same species. Length 30 mm.

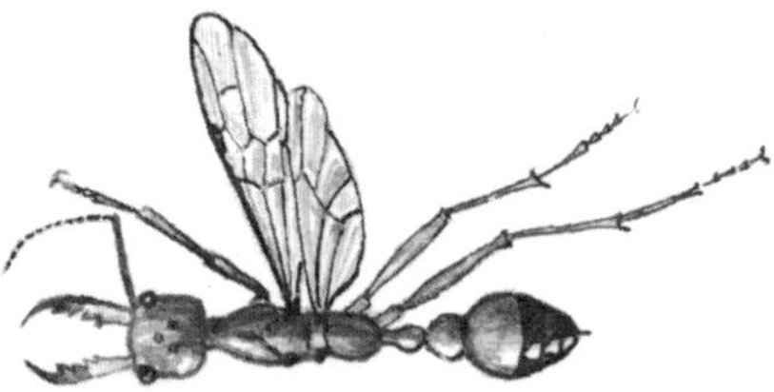

Female, same species. Length 20 mm.

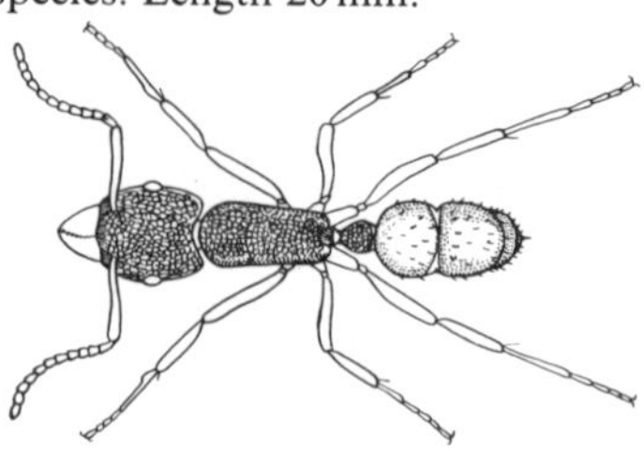

The green-head ant, a 'jumper' of the *Chalcoponera* genus.

actual eggs are much smaller and white, resembling castor sugar. Adults feed on nectar and sugar but larvae are fed on insects. The most primitive ants in the world are in the genus *Nothomyrmecia* found in Western Australia. The species *Myrmecia mjobergi* nests in the epiphytic ferns.

FIRE ANTS (Subfamily MYRMICINAE)

Ants in the species *Pheidole megacephala* have large heads and are known as 'fire ants'. A related species, *P. gratiosa*, also has a large head but is reddish in colour. Species of *Pheidole* often inhabit the hollow tubers of the ant house plants *(Myrmecodia* and *Hydnophytum)* which grow on the trunks and branches of mangroves and other trees. These bulbous plants house hosts of ants which, when the stem is tapped, will swarm over the surface and attack an intruder.

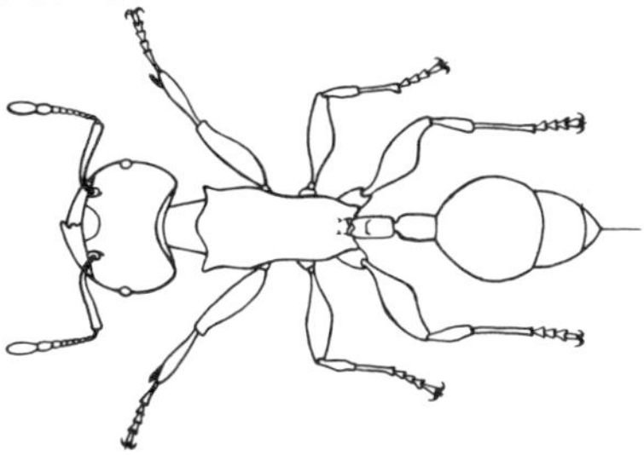

The reddish *P. gratiosa*. Length 9 mm.

Also included in this subfamily are the minute red meat-eating or hospital ants, also known as 'Pharaoh's ant' *(Monomorium pharaonis)*, and the Singapore ant *(M. destructor)* Ants with heart-shaped abdomens are included in the genus *Crematogaster*. Adults in this subfamily have stings. Their pupae are not enclosed in cocoons and lie uncovered in the nest.

STINGLESS ANTS (Subfamily DOLICHODERINAE)

Possibly the best-known species is the ubiquitous meat or mound ant *(Iridomyrmex purpureus)* whose huge, gravel-capped nests are frequently seen in the bush. These ants care for ant-cows (lerps or other honeydew-producing insects) which they 'milk' for food. In times of drought, they may also attack dead stock.

Pests in this group include the tiny white-footed black house ant *(Technomyrmex albipes)* which is attracted by sugar and other sweet foods found in homes, and the introduced yellow-brown Argentine ant *(Iridomyrmex humilis)* which is a pest in various areas. In summer they may infest houses unless controlled by insecticides.

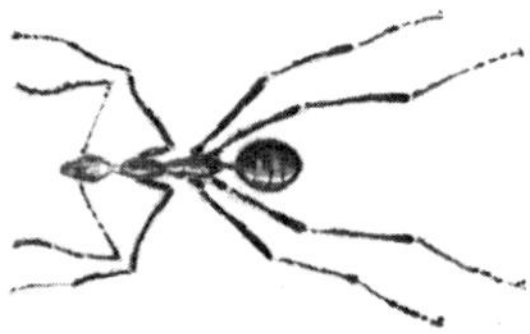

Common stingless ant.

GREEN TREE ANTS, HONEYPOTS and others (Subfamily FORMICINAE)

Also lacking stings, the ants of this subfamily have a small aperture surrounded by minute hairs at the end of the abdomen. This acts as a nozzle, through which defensive formic acid is released at intruders. Possibly the best-known are the green tree ants *(Oecophylla smaragdina)*. These ants act as a communal group and build nests of leaves of tropical trees by drawing them together and using their larvae to produce silk, with which the leaves are bound together. Worker ants form long chains to drag leaves spectacular distances and their co-operation in nest building is one of the wonders of the insect world. The end worker grabs the leaf and in turn the worker is grasped round the waist by another worker until 50 or 100 ants are linked in a single chain. Once a nest has been built it should not be disturbed as the ants are most aggressive and will swarm at any intruder, inflicting painful bites which they then spray with formic acid. When defending the nest, the green tree ant rears on its hind-legs and menaces the intruder with open jaws.

Honeypot ants *(Camponotus inflatus, Melophorus bagoti, M. cowleyi)* live in arid areas. To provide a store of honeydew and nectar, certain workers *(repletes)* have abdomens which can become distended to the size of a small grape. These provide a storehouse for nectar and honeydew from which other ants draw food when needed. Desert aborigines still excavate honeypot ants' nests and eat them for their sugar content. When filled with nectar, the storage ants cannot move, and remain clinging to the roof of the underground nests in clusters. When a replete has been exhausted, its abdomen contracts to normal size and the insect resorts to normal nest activities.

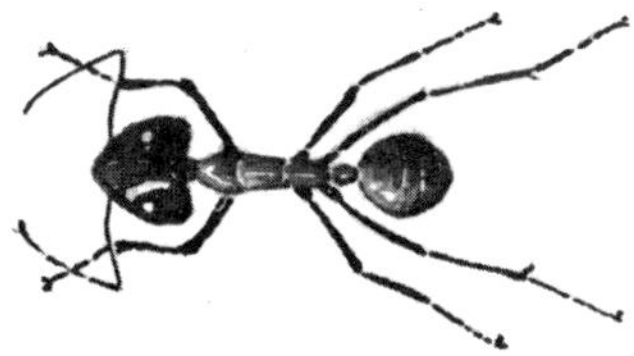

Camponotus worker, abdomen undistended.

The golden ant *(Polyrhachis ammon)* has strong spines on its thorax and beautiful golden metallic colours. While some species in the genus *Polyrhachis* excavate underground nests, others build nests of silk and plant fibres on leaves, twigs and the bark of trees.

Aquatic Insects

Mayfly larvae

Order EPHEMEROPTERA

Many different forms of mayfly larvae are found in Australian streams and other aquatic locations; all are conspicuous for the triple 'tails' *(cerci)* which protrude from the rear of the abdomen. The larvae also possess rows of hairs or leaf-like structures on both sides of the body. These are the tracheal gills, which enable the larvae to obtain oxygen underwater.

After mating, which sometimes takes place in mid-air, the female mayfly deposits her eggs directly into the water. The specific technique differs between species but on contact with the water, the egg mass breaks up and sinks to the bed, where individual eggs are fixed to pebbles with tiny filaments.

In some species the female retains the fertilised eggs for some time. When they are dropped into the water, the embryos hatch in a few minutes. Otherwise, hatching takes from seven to 14 days and the nymphs which emerge have a fairly long underwater life, averaging about 12 months. With some larger species, nymphs may remain aquatic for two years, with up to 27 moults.

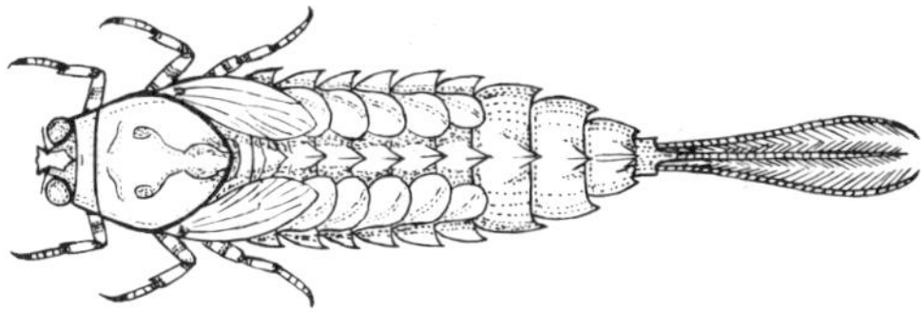

Mayfly nymph of *Oniscigaster* genus. Length 25 mm.

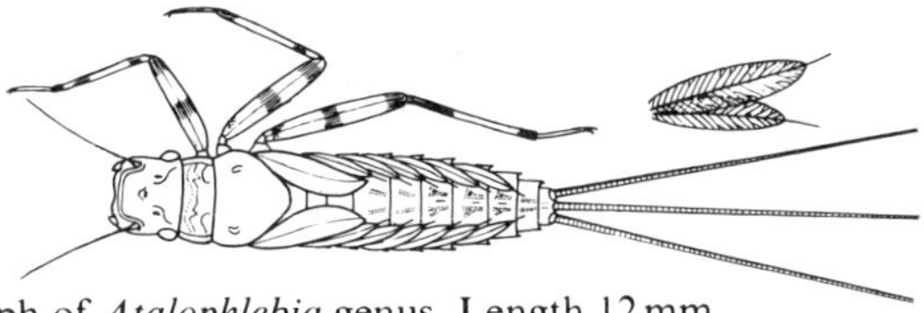

Mayfly nymph of *Atalophlebia* genus. Length 12 mm.

Mayfly nymphs are very active, feeding continuously on decaying plant material or on minute freshwater organisms which they catch with their long, sharp mandibles. Others are sedentary and burrow into sand or silt where they live on the detritus they excavate. In free-swimming mayfly nymphs, the cerci are often covered with hairs; by rapidly flicking their abdomens, the tails straighten and act as a paddle to propel the nymph through the water.

When the time comes for the nymph to be transformed into an adult, it climbs onto a rock and moults and becomes a *subimago*. In this intermediate stage, the mayfly has wings and the insect will remain in this state from anything between five minutes to a day. In this respect the mayfly is unique in that it is the only insect to undergo a final moult after acquiring functional wings. Depending on temperature and humidity the final moult occurs and the mayflies then swarm. Their short adult life is spent in mating—they die sometimes only a few hours later; rarely do they live more than a week.

Mayfly larvae are eaten by freshwater fish and are also attacked by carnivorous dragonfly nymphs and other aquatic insects. Fishermen often use lures based on the mayfly larva's appearance.

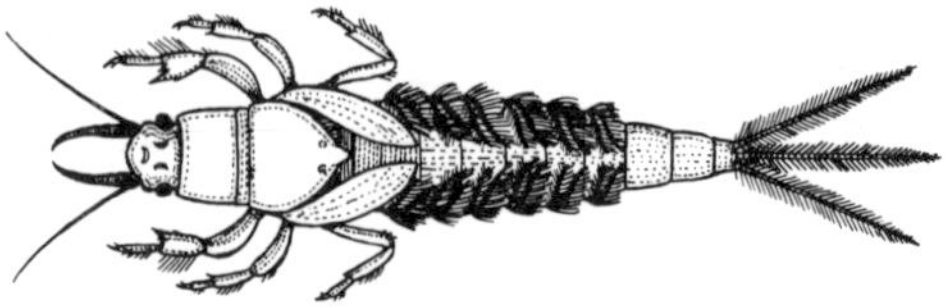

Mayfly nymph of *Ichthybotus* genus. Length 27 mm.

Damselfly and dragonfly larvae

Order ODONATA

From the time the eggs hatch, nymphs of damselflies and dragonflies live wholly underwater inhabiting many different locations, including sluggish freshwater rivers, creeks and swamps; some species may be found in slightly salty inland waters; others in rapidly flowing or even torrential streams.

Although capable of movement, they are not active, although they are adept hunters, using stealth and patience to satisfy their voracious appetites.

Damselfly nymphs exist underwater using three (rarely two) leaf-shaped gill plates which extend from the tail end of the abdomen. These extract oxygen from the water and may also be waved to propel the nymph through the water. The shape of the gill plates is used as a means of identification of different groups.

In contrast, dragonfly nymphs suck water into a chamber lined with oxygen-extracting gills. However, when a dragonfly nymph needs to move quickly, it can expel this water with a sharp muscular contraction and this 'jet propels' the nymph a considerable distance much more quickly than it could move with its legs.

The dragonfly nymph is longer and heavier than the damselfly although both possess an extraordinary development of the lower lip *(labium)* known as a 'mask'. This has become strangely modified into a grasping organ—a large triangular-shaped plate hinged beneath the jaw. At its farthest

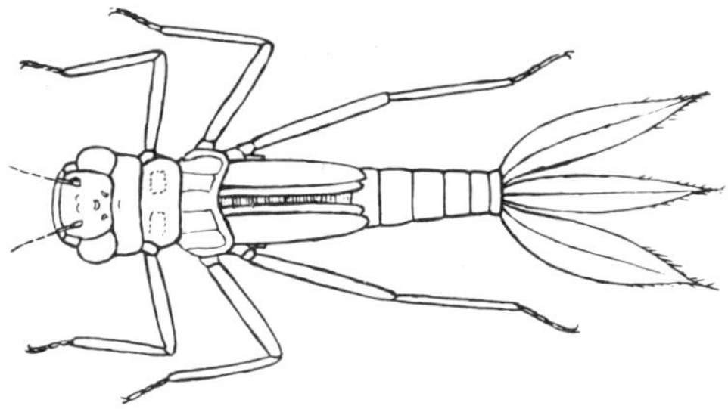

Larva of damselfly, *Diphlebia lestoides.*

end, there are two palps and these are armed with sharp spines and teeth which differ in shape between species. This 'mask' may be moved rapidly and when hunting the mask remains hidden beneath the head until the nymph comes within reach of its prey. Then the mask is projected forward and the sharp spines impale the victim, which may be a mosquito larva, fish fry or a tadpole. The food is then carried in the mask to the mouth where strong jaws consume it.

Dragonfly nymphs are aggressive cannibals and in turn are eaten by large fish, frogs, birds and reptiles; other large aquatic insect larvae may also eat them.

Throughout their development, nymphs of both suborders are active and always hungry. From the time they emerge from the egg, damselfly and dragonfly nymphs pass through 10 to 15 moults, the number varying between species and possibly affected by water temperature.

The life-span of an Odonata nymph is variable. In locations subject to rapid summer evaporation, nymphs may mature in 10 to 15 weeks. In cooler areas with permanent water, nymphs may remain in this stage of their life-cycle for several years, although 6 to 18 months is said to be most usual.

When the time comes for transformation from nymph to adult, the nymph leaves the water and crawls away from the edge. There the final skin is shed and the slender, long-winged insect takes to the air (see also damselflies and dragonflies, pp. 5–7).

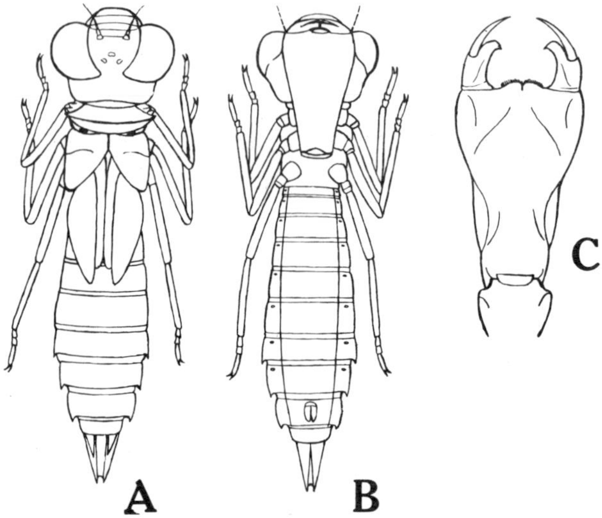

A—Dorsal view of larva of dragonfly, *Aeschna brevistyla*. Length 45 mm. B—Ventral view of same, mask retracted. C—extended mask.

Stonefly nymphs

Order PLECOPTERA

Stonefly nymphs are conspicuous for their long antennae and cerci, which may be of the same length. They are wholly aquatic and vary in length from 10 to 50mm. Legs are well developed and the head is quite distinct. Gills may be located in various parts of the body and identification is often based on the position and form of the gills.

To survive, stonefly nymphs need clear, well-oxygenated water, usually where there is a gravel bottom and a fairly fast water flow. The nymph may pass through 22 to 33 moults and the nymphal stage may last from one to three years. As an adaptation to the current of fast-flowing streams, stonefly larvae often have flattened bodies.

The mouthparts of all stonefly larvae are of the biting type and although many are carnivorous, those inhabiting

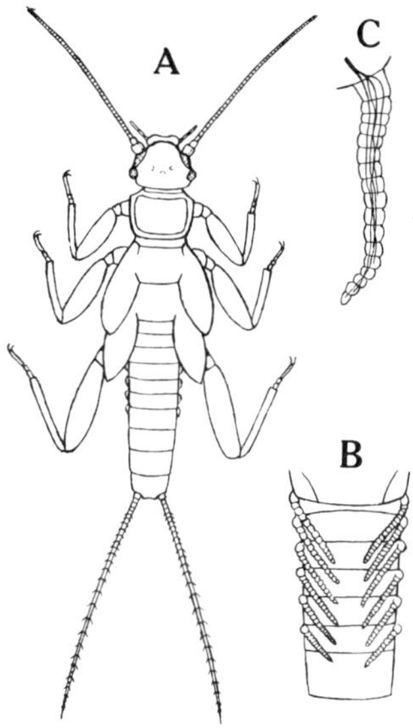

A—Larva of stonefly, *Eusthenia spectabilis.* Length 28 mm. B—Magnified ventral view of abdominal gills of same. C—Single gill.

cooler waters in the suborder Filipalpia are said to be vegetarian.

Stonefly nymphs may be found under stones in clear, fast-flowing streams though occasionally they may also be collected from heavily vegetated and turbid waters.

Water bugs and their relatives

Order HEMIPTERA Suborder HETEROPTERA

Fifteen families have been described among the Hemiptera which are closely associated with water. Of these, only the following are included as the others are very small, rarely reaching an adult size exceeding 3mm.

The grouping of aquatic bugs varies between different orders and that shown below is based on Williams (1968) with the actual order based on Woodward, Evans and Eastop (1970). The grouping is based on the ecology and habitat of the group and is therefore of most value as a guide to identification.

Group	*Common name*	*Family*
Shore-living semi-aquatic	Shore bugs	SALDIDAE (see Bugs, p. 66)
	Toad or sand bugs	GELASTOCORIDAE
Restricted to water surface	Pond skaters and water striders	GERRIDAE
	Water measurers	HYDROMETRIDAE
Underwater living but mostly air breathing	Back swimmers (sometimes wrongly called 'water boatmen')	NOTONECTIDAE
	Water scorpions	NEPIDAE
	Giant water bugs	BELOSTOMATIDAE
	True water boatmen	CORIXIDAE

Like their terrestrial relatives (see Hemiptera pp. 49–69) these aquatic bugs undergo metamorphosis with about five moults occurring between juvenile and mature forms. Often they are wingless and lengths range from microscopic to several inches. Most aquatic bugs are carnivorous and feed by sucking body fluids from the small aquatic fauna which they catch. They may also eat insects which fall onto the water surface.

TOAD BUGS (Family GELASTOCORIDAE)

Toad bugs of the inland-dwelling type often occur among fallen leaves on forest floors where they will burrow into the soil. Others, preferring proximity to water and camouflaged by their drab colouring, live on mud or sand-banks, sometimes crawling beneath the surface of the water.

Toad bugs grow to about 7mm in length and in appearance they are somewhat square, with their backs roughened and resembling the warts of toads. The resemblance continues with bulging eyes and their habit of hopping as they search for prey, which consists of small flies and other small insects which are seized by the two front legs. Some species deposit eggs in burrows dug in wet sand. When not hunting, adults retreat to burrows. The family name Gelastocoridae is based on the Greek 'laughable bug'.

Australia's 20 species of this insect are all grouped in the genus *Nerthra*.

POND SKATERS and **WATER STRIDERS** (Family GERRIDAE)

Long-legged, active insects commonly seen skimming along the surface of ponds, streams and small dams. Often they are in small groups which scatter when approached. While most species in this family are confined to inland waters, some are found around the coast and live on brackish or sea water. Overseas, one species has been seen in the open sea hundreds of kilometres from land. Marine species have much shorter abdomens than freshwater forms. Typical are the species *Halobates whiteleggi* and *Gerris australis* both of which may be seen on Sydney harbour. Adult males are

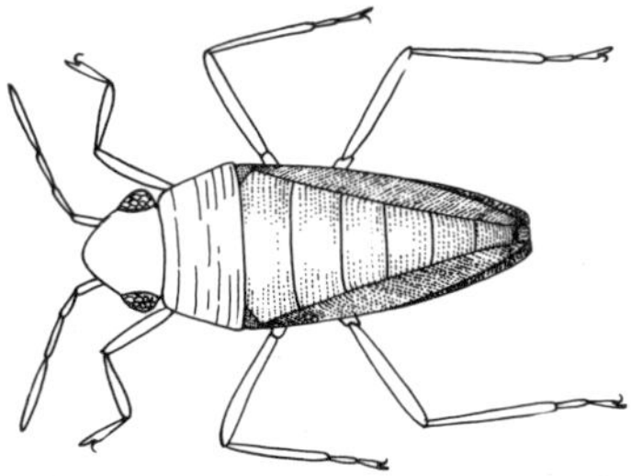

Water strider.

found in both winged and wingless forms and the latter make interesting subjects for study in an aquarium. However, they are easily drowned and should be carried in wet moss rather than water.

Movement on the water is by means of microscopic feathery hairs on the feet and progress is by gliding across the surface of the water. Even when the insects leap into the air, they never break the water surface and leave only the slightest dimple.

The most common species is the Australian water strider *(Gerris australis)*.

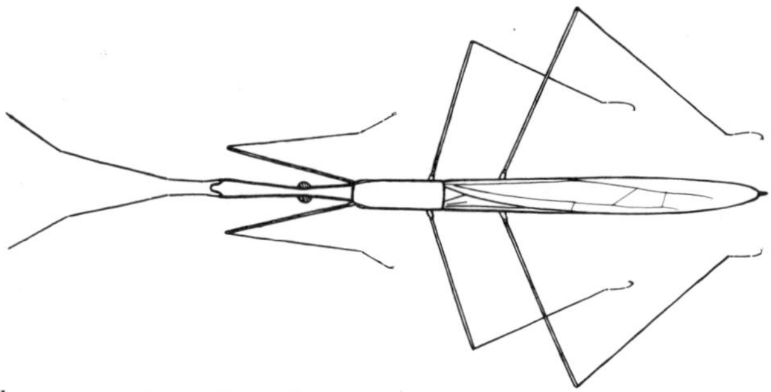

Hydrometra strigosa. Length 11 mm.

WATER MEASURERS (Family HYDROMETRIDAE)

Less conspicuous than water striders and slower in their movements, adults are about 10 to 15mm long and are more elongated and slender than the water striders. They may often be collected by dragging a net among reeds and other vegetation in shallow water.

BACK SWIMMERS (Family NOTONECTIDAE)

Considered to be the bugs best adapted to an aquatic life, nymphs live wholly underwater and adults only near the surface to obtain oxygen through breathing siphons in the tail. The abdomen has a hair-covered groove on each side and in this adults trap air to allow them to remain underwater.

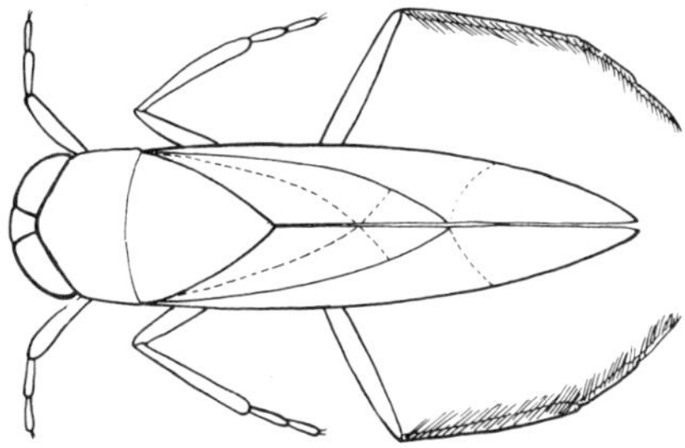

Back swimmer.

The back swimmers are easily recognised by their convex dorsal surface which resembles the keel of a boat. They swim upside-down using their elongated hind-legs, which are broad and feathered with short stiff hairs. These can propel the insect through the water at a fast speed. Prey is caught with the slender front legs and sucked by the beak.

While most aquatic insects are darker above than below, the colouring of back swimmers is the opposite; the back may be almost white while the abdomen can be very dark green or brown.

Back swimmers also make interesting aquarium subjects and can be taught to take a fly held at water surface. Meat may be used as a substitute but it must be held in forceps and moved about to simulate live food. Back swimmers can fly and aquaria must be covered with gauze to prevent them escaping. When handling avoid a jab from the beak which can be painful.

WATER SCORPIONS (Family NEPIDAE)

These are all elongated insects. Some are thin and stick-like while others have a broader, flattened body. All are conspicuous for two long half-breathing tubes which extend like a tail from the end of the abdomen and which are joined when air needs to be obtained. Some stick-like water scor-

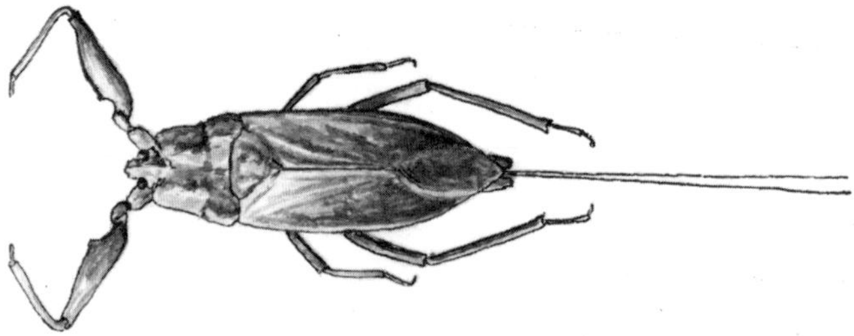

Water scorpion of *Laccotrephes* genus.

pions in the genus *Ranatra* may be as much as 50mm long and have a breathing tube which is equally long. The broader forms vary from 35mm in length downwards and have varying lengths of breathing tubes.

Water scorpions live in still or slowly-flowing fresh water and may often be found in farm dams. Their movement is sluggish and ungainly although food is obtained by surprisingly quick grasping movements of the front legs. After catching its prey, the insect injects poisonous saliva into it through the beak, which is then used to drain the body fluids. Prey may include insects, tadpoles and small fish. Water scorpions may often rest for long periods on a submerged object with only the breathing tube penetrating the surface. Most species have wings and may fly extended distances to find fresh water if their home territory evaporates or if food becomes scarce.

In some places, water scorpions are called 'toe biters' although McKeown reported that he never obtained evidence that this name was justified.

Water scorpions deposit oval eggs on the stems of aquatic plants, each egg bearing a tuft of white hairs at one end.

GIANT WATER BUGS (Family BELOSTOMATIDAE)

Often misnamed 'water beetles', these are possibly the most aggressive of all freshwater insects. Overseas they may grow to a length of 150mm and have been known to capture and kill fish nearly twice this size.

Giant water bugs propel themselves with strong strokes of the middle and hind-legs, which are flattened and feathered. They need to surface for air about once every 30 to 60 minutes; air is stored in a hair-edged space under the top of the abdomen and below the wings. At rest, giant water bugs can remain submerged for prolonged periods.

Of special interest is the 20mm long bug *Diplonychus rusticus*. The eggs of this water bug are deposited by the female on the male's back, probably to protect them from predators. They are carried by him until the young hatch, although empty cases may often adhere long after the young have emerged. Nymphs are wholly aquatic. In South America and in Indonesia, larger species are eaten raw or baked after the legs and wings have been removed.

Giant water bugs often migrate between ponds and they may be attracted to street and house lights in large numbers mistaking the illuminated area for moonlit water. When in a lighted area, they thresh violently against walls and ceilings in search of water.

Australia's largest species *Lethocerus insulanus* attains a length of 70mm and is a rich brown. It will kill insects, fish and frogs, capturing them with its forelegs and killing with poison injected by the strong sharp beak. This beak can inflict a painful injury.

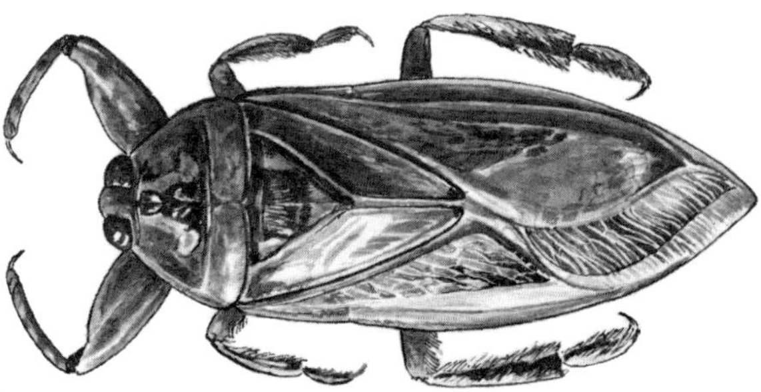

Lethocerus insulanus. Length 70 mm.

TRUE or **LESSER WATER BOATMEN** (Family CORIXIDAE)

Mainly plant suckers, these insects have modified forelegs which they use to gather algae and other organic matter from the pond bottom. Some species eat mosquito larvae. Only the hind-legs are broadened and bristly for swimming; the centre pair are used mainly for clinging. Water boatmen swim with the wing-case uppermost and lack the back swimmer's long hind-legs.

Adults rarely seem to grow larger than 15mm in length and while colours are dull, surfaces may often be patterned. They are found in vegetation or on muddy pond bottoms, usually close to the edge of large areas of water. Air may be stored beneath the wings over most of the body surface although adults can emerge from the water and are strong fliers.

Water boatmen are good specimens to house in aquariums. They should be given rocks and water plants where they may emerge, but the aquarium should be covered to prevent escape. They may emerge at night and stridulate by rubbing pegs on the forelegs against the head to produce a short chirping sound.

In Mexico, the eggs of one species are collected and dried by natives who then use them as food, either baked, as a dressing for meat or added to stews.

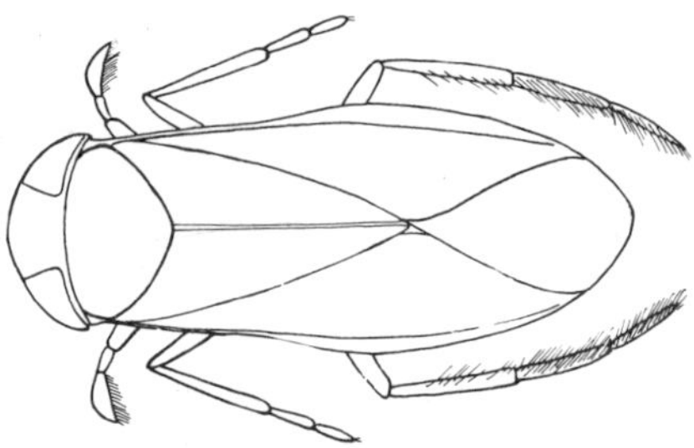

Water boatman.

Alderfly and Dobson fly nymphs

Order MEGALOPTERA

Alderfly larvae grow to more than 24mm in length and appear to have ten pairs of hairy legs and a single, hairy, pointed tail. Only three pairs are genuine legs, the other seven pairs being abdominal gills.

In contrast, the larvae of the genus *Archichauliodes*, a Dobson fly, have eight pairs of abdominal gills and these lack the 'hairs' found on alderfly larvae. Also instead of the single, long, hairy tail of the alderfly, Dobson fly nymphs have two much shorter rear abdominal appendages or pro-legs. Some Dobson fly nymphs may grow to a length of 50mm.

The eggs of Megaloptera are laid on rocks and branches overhanging the water. The newly hatched larvae fall into the water and sink. On the bed they hide beneath rocks or in other suitable retreats. The development rate and number of moults is not known but it has been deduced that their underwater life may last two years, at least.

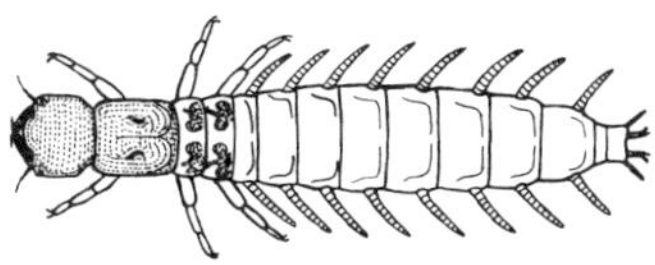

Larva of *Archichauliodes* genus. Length 27 mm.

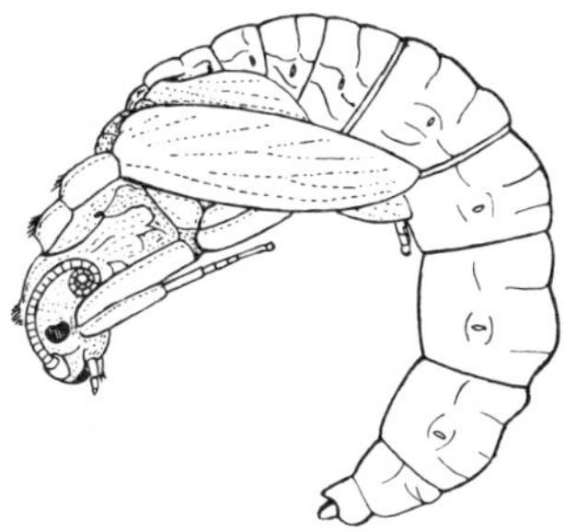

Pupa of same.

Larvae are active and carnivorous and live on other underwater life until they attain maturity. They then move to the edges of the streams when waters are high. When the water recedes, they are stranded ashore, usually beneath stones and logs or buried in the soil. Each larva makes a pupal cell and pupation lasts from two to four weeks. Throughout this period, the insect's legs are free and it can move about, using its jaws or mandibles to defend itself against predators.

Water beetles

Order COLEOPTERA

Fourteen families of beetles in Australia spend all or part of their lives associated with water. In some, both larvae and adults are fully aquatic though others are aquatic only during part of their life-cycle.

Some of these beetles may have an adult length which is less than 8mm and most of these smaller species have been omitted here. The families included are:

Water tigers	DYSTICIDAE
Whirligig beetles	GYRINIDAE
Water scavenger beetles	HYDROPHILIDAE
Riffle beetles or water pennies	PSEPHENIDAE

GUIDE TO IDENTIFICATION OF ADULTS

Water-surface swimmers, skimming rapidly when disturbed, often in large numbers . . .**whirligig beetles**

Underwater swimmers, maxillary palps have bristly club and are nearly always longer than antennae. . .**water scavengers**

Underwater swimmers with eyes recessed and prothorax presenting a smooth outline; cups on front feet of males . . .**water tigers**

GUIDE TO IDENTIFICATION OF LARVAE

Six-segmented legs with last three segments well developed; elongated body ending with abdominal segment bearing four hooks, lateral gills on each abdominal segment . . .**whirligig larvae**

Last abdominal segment lacking hooks but often with two short cerci; prominent sickle-shaped jaws

. . . **water tiger larvae**

Body flattened and wide, an elongated oval when viewed from above, adheres to submerged rocks in rapid streams

. . . **water penny larvae**

Maxillary palps usually more noticeable than antennae which are shorter; body tapered and sometimes flattened with no spines beneath last segment of abdomen. Small gills on each abdominal segment and two small cerci at tip of abdomen

. . . **water scavenger larvae**

WATER TIGERS or **CARNIVOROUS DIVING BEETLES** (Family DYSTICIDAE)

With its streamlined body and hind-legs which are hair-fringed and act as oars, the water tiger is well adapted to its environment. The antennae are long and threadlike *(filiform)*. Air is obtained from the atmosphere and stored between the elytra and the top of the abdomen; when a fresh supply is needed, the beetle rises to the surface, thrusts the tip of its abdomen above the surface and, by raising the elytra, takes in fresh air.

Water tigers are fierce predators of all living animals they can catch. Prey includes aquatic insects, fish and amphibian spawn.

Water tiger larvae are also carnivorous and air breathers. They float on the surface with the tail protruding. They do not, however, eat the flesh of their prey as do the adults. Capture is by sharp extended inward-curving jaws *(mandibles)* which are hollow. Through these, digestive juices are

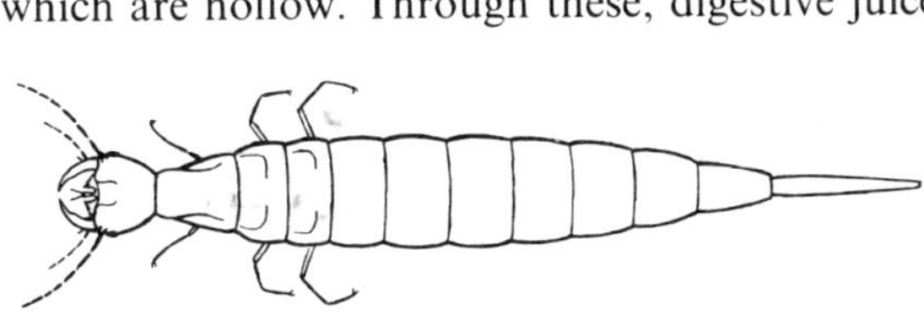

Dytiscid larva.

pumped into the prey and in time, the body juices and digested body tissues are ingested by the larvae. The larvae's legs are well adapted for swimming; they pupate in clay above the water level in ponds or in stream banks.

There are 188 described Australian species, adult lengths ranging from 5 to 25mm. Australian species have a wide colour range, from black and dark brown to olive green with yellow margins. This latter colouring is found on Australia's two largest species *Cybister tripunctatus* and *Homeodytes scutellaris*. The most common species *Eretes sticticus* is yellowish-brown and about 12mm in length. Males in the genera *Cybister* and *Eretes* have suction cups on the forelegs which are used to grasp the female when mating.

Adults fly at night between expanses of water and may be diverted by house or street lights. On land they are awkward, pushing themselves along using only their large hindlegs.

Large dysticids are used for food in China, where they are collected and dried before being eaten uncooked.

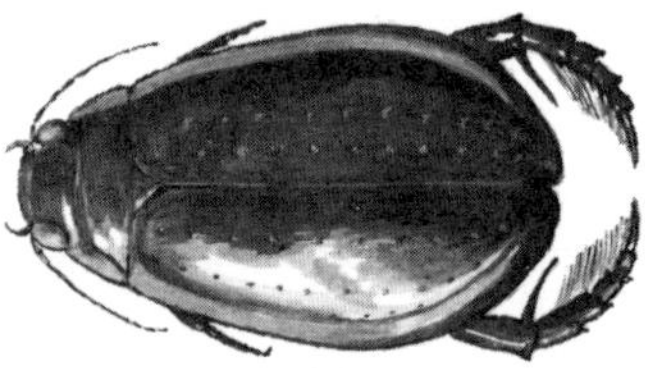

Water tiger of *Cybister* genus. Length 40 mm.

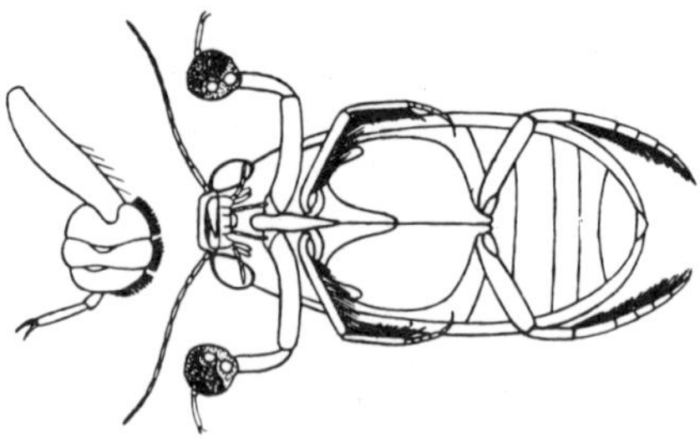

Ventral view of *Eretes* male showing suction cups. Also, dorsal view of suction cup.

WHIRLIGIG BEETLES (Family GYRINIDAE)

Whirligig beetles have a smooth, flattened and rounded body, short thick antennae, and hind and middle legs that are modified into short, broad paddles. The forelegs are long and used for grasping. The tendency of these beetles to swim on the surface, often in large numbers, makes their presence easily detected. Each eye is separated into two parts by the side margins of the head. One pair is used for vision above the water surface and the other provides underwater observation at the same time.

One of the most common species in eastern Australia is *Macrogyrus oblongus*, a dull green insect with a noticeably grooved elytra. The largest Australian species, found in the mountains of New South Wales, is *Macrogyrus striolatus*; this may be dark blue or olive green in colour and have a length of up to 17mm. Altogether, 28 species have been described in Australia, compared with the known world population of some 700 species.

Whirligig beetles are found both in lakes and streams and swim in circles. Adults have a special organ in the antennae which is sensitive to water surface movement and this helps them to locate insects and other food.

Eggs are deposited on stems of aquatic plants and those of the species *M. strigosus* are often found on growing rice in the Murrumbidgee Irrigation Area, New South Wales. Larvae are wholly aquatic, obtaining oxygen through ten pairs of feathery gills. They are long and slender with sickle-shaped jaws which they use to catch their prey. After their second moult, larvae leave the water to pupate in a dark, irregularly oval cocoon, usually attached to a plant a short distance above the water level.

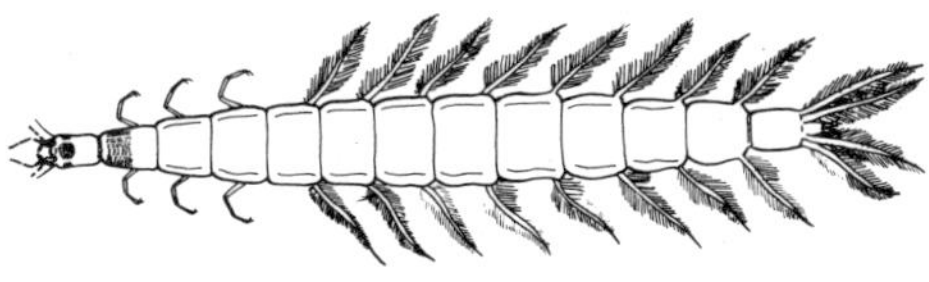

Larva of *Macrogyrus* genus. Length up to 20 mm.

WATER SCAVENGERS (Family HYDROPHILIDAE)

Ranging in length from 3 to 35mm water scavenger beetles have a smooth, oval shape and are a greenish black. They may be distinguished from tiger and whirligig beetles by their maxillary palps, which are longer than the clubbed antennae and often of distinctive shape. Palps are used to break the surface film of the water when the beetles ascend for air, which is stored between the elytra and the abdomen.

Their bodies are deeper than those of the whirligig and when swimming, hydrophilids use the legs alternately not together as do water tigers. This action gives water scavenger beetles a strangely erratic course underwater compared to the straight dive of water tigers.

Of the 80 species described from Australia, those in the genus *Hydrophilus* often have a noticeable 'keel' on their under-surface. The largest, *H. latipalpus*, may attain 40mm in length.

Eggs are deposited in capsules or silken cocoons made by the female and either carried by her or attached to aquatic plants. Larvae are slender and flattened with short, well-developed jaws *(mandibles)* which they use to obtain live food or carrion. These larvae have less conspicuous abdominal gills than those of whirligig larvae.

Water scavengers are found in both still and running water and one species *(Enochrus andersoni)* has been found in salt lakes in Victoria.

Hydrophilus latipalpus.

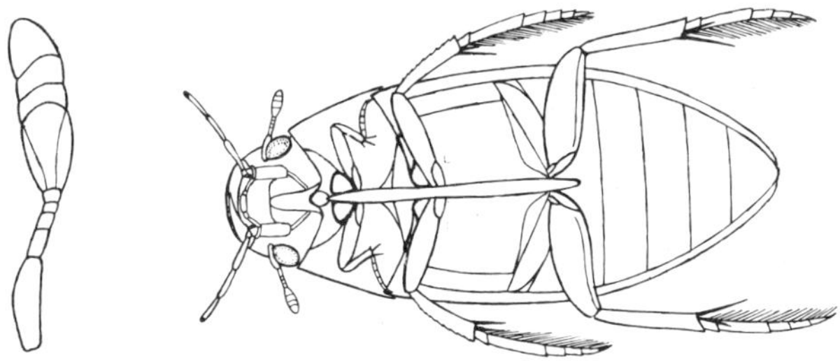

Water scavenger, *Hydrophilus albipes*. Also, antenna. Length 37 mm.

Adults are considered to be vegetarians although the introduced species *Cercyon haemorrhoidalis*, with red-marked elytra, lives in rotting vegetation, cow manure and earth which is constantly moist.

Another widespread species is *Hydrophilus ruficollis* which grows to about 25mm in length and is olive black. The species *Sternolophus marginicollis* is small, with a shiny body.

RIFFLE BEETLES or **WATER PENNIES** (Family PSEPHENIDAE)

Best recognised by their flattened, disc-shaped larvae which cling to stones in swiftly running streams, the 'water pennies' are well-known in mountain streams in Victoria and in the inland waters of New South Wales and Tasmania. The larvae seem to live mostly where the water-flow is fastest, on a surface where there is chiefly gravel and small stones.

Larvae are small, not more than 8mm in diameter, and the whole body is used as a sucker which is so effective that to dislodge them from stones takes considerable skill. When dislodged, the larvae tend to curl up like chitons.

Adults of only one species, *Sclerocyphon maculatus*, have been described in Australia. They are about 5mm in length, oval and flattened and black or dark brown. They are covered with a faint haze of hairs or bristles. The legs are fine and the head can only be seen from above.

Aquatic larvae of true flies

Order DIPTERA

Although some 13 families of Diptera have aquatic larvae, not all larvae in each of these families are aquatic; for example, crane flies have both aquatic and terrestrial larvae. Only six of the more important aquatic families are described here.

Aquatic larvae of the Diptera are usually worm-like, with up to 12 body segments. They have no jointed legs or compound eyes. Heads vary in size but antennae are usually short. Prolegs are sometimes present but if they are, they are short, stumpy and without joints. Some larvae have special organs enabling them to obtain air although many are able to obtain oxygen directly from the water and remain submerged.

Eggs may be deposited in different ways, ranging from random scattering on the water surface to attachment in masses to submerged objects. Larvae usually moult about three times and, excepting those of mosquitoes and midges, pupae are inactive.

CRANE FLY LARVAE (Family TIPULIDAE)

Varying in length and sometimes reaching 50mm, crane fly larvae have a spiracular disc at the back of their body which is usually lobed; there may also be anal gills. Larvae are usually white or brown and some have been found in salt lakes. Identification of groups is normally based on the number of lobes on the spiracular disc.

MOTH FLY LARVAE (Family PSYCHODIDAE)

Found in decomposing organic matter (on which they feed) usually at the edges of ponds, polluted farm dams and stagnant waters, these 12mm long worm-like creatures are identified by the tendency for segments to have secondary divisions. The hindmost segments have hard areas of skin on the upper surface. There are no prolegs.

MOSQUITO LARVAE (Family CULICIDAE)

All mosquito larvae and pupae are aquatic and larvae pass through four moults. During their larval phase they eat detritus or plant matter. The larval period is dependent on temperature and may vary from a few days to several months. When disturbed, larvae dive with snake-like movements which have earned them the name 'wrigglers'.

The 'comma' shaped pupae are active and swim quickly by means of paddle-like tails. They have a bulbous cephalothorax and a narrow abdomen. Two breathing trumpets project from the cephalothorax and their shape may often determine the tribe. Pupae remain in this stage from a few days to a maximum of about three weeks. When pupation is complete, the pupa floats on the surface and the perfect insect splits the skin along the back of the pupal case, emerges and rests upon the surface of the water until its wings have hardened.

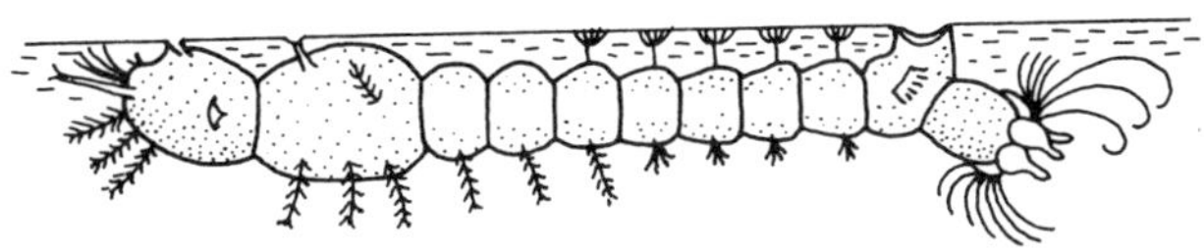

Aquatic stage of mosquitoes.

The majority of Australian mosquito species are grouped into two main tribes—the Anophelini and the Culicini.

Eggs of the Anophelini have floats and when larvae hatch, they are easily recognised by the way they lie suspended horizontally in the water, often close to the surface. The pupae have short breathing trumpets with a wide opening.

Eggs of the Culicini do not have floats and the larvae, which have a breathing siphon at the tail, usually rest in the water at an angle of about 45 degrees to the surface. The pupal breathing trumpets are either conical or elongated.

BLOODWORMS or **TRUE MIDGE LARVAE** (Family CHIRONOMIDAE)

From eggs deposited in the form of long, gelatinous ropes or in sausage-like masses, emerge the slender 'bloodworms' which are the larvae of true midges. As their name suggests, these larvae are bright red and worm-like, with a distinct head and a pair of prolegs on both the first thoracic and last abdominal segments. Larvae are mainly herbivorous, although some are carnivorous. Not all are red; larvae may be white, green or yellow.

Pupae are very similar to those of mosquitoes and some are similarly active. However other species fasten themselves to submerged surfaces and make a cocoon. These only become active when pupation is ended, when they rise to the surface so that the adult may emerge, usually on a warm summer evening. Many fish rely on bloodworms as a major source of food.

NET-VEINED MIDGE LARVAE (Family BLEPHARICERIDAE)

Although adult net-veined midges resemble mosquitoes, the larvae are unique. They are usually less than 13mm in length, and flattened. The body is in six divisions, each separated by a waist; on the underside of each division is a sucker. These suckers enable the larvae to cling to rocks in mountain waterfalls and in fast-running streams as they feed on algae. The dark, broadly oval pupae possess suckers on their sides; they breathe through comb-like structures in the front of the body.

Net-veined midge larva.

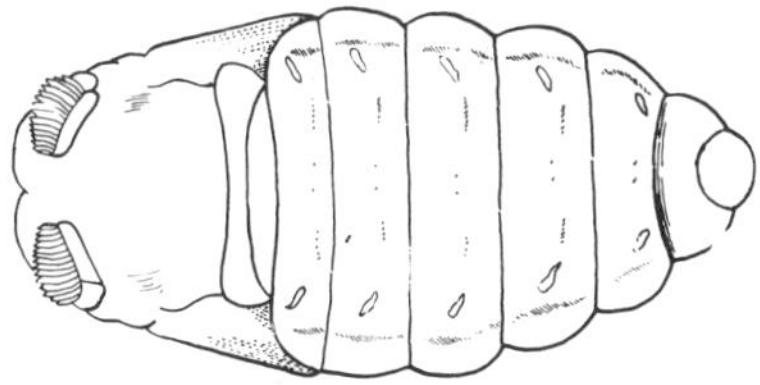

Net-veined midge pupa.

RAT-TAILED MAGGOTS (Family SYRPHIDAE)

Larvae of the introduced drone fly *(Eristalis tenax)* and other species in this genus are known as 'rat-tailed maggots' because of the telescopic 'snorkel' at the back of the body.

Larvae breed in drains, sewage pools, polluted ponds and other stagnant water, and have even been found in the remains of decaying animals. This latter occurrence has given rise to a myth that bees sometimes breed in carcasses. Although the larvae live on decaying organic matter, they must breathe air; for this, the breathing tube is extended or contracted depending on the depth of the liquid in which they are feeding. Their pupa is the final larval skin and this often has the remains of the distinctive 'rat-tail' clearly visible.

Water 'case moths' or caddis fly larvae

Order TRICHOPTERA

Adult caddis flies lay long strings of eggs enclosed in a jelly-like mass or string, which they attach to plants overhanging the water or lay directly into water. Females of some species whose larvae do not make cases, enter the water and attach the egg mass to a submerged rock or plant. One genus, *Plectrotarsus*, has the egg mass camouflaged by hairs from the under-surface of the female's wings.

Caddis fly larvae are best known for their distinctive 'cases'. All live in water and without their case, resemble

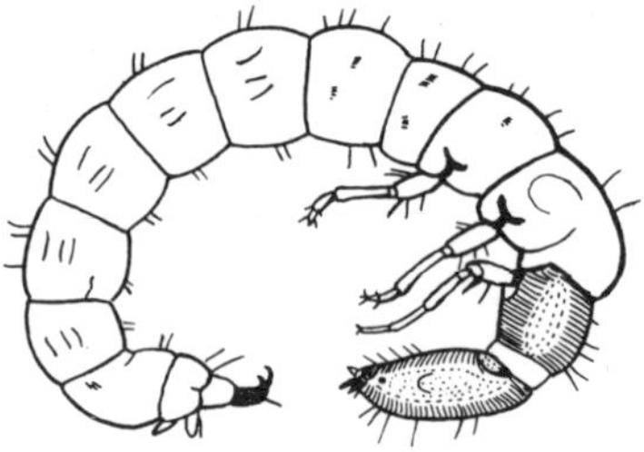

Caddis fly larva of *Helicopsyche* genus. Length 10 mm.

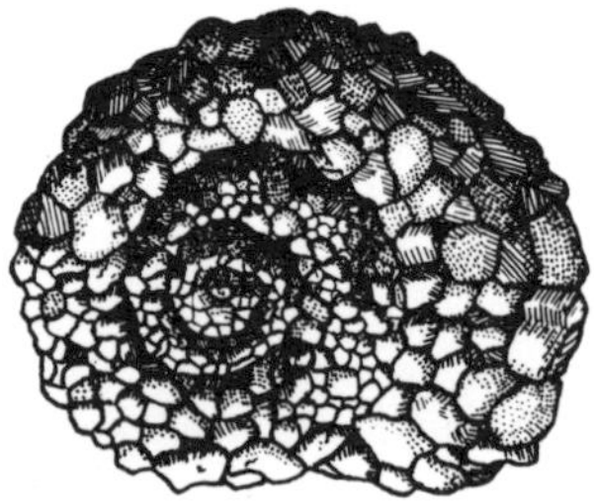

Portable case of same.

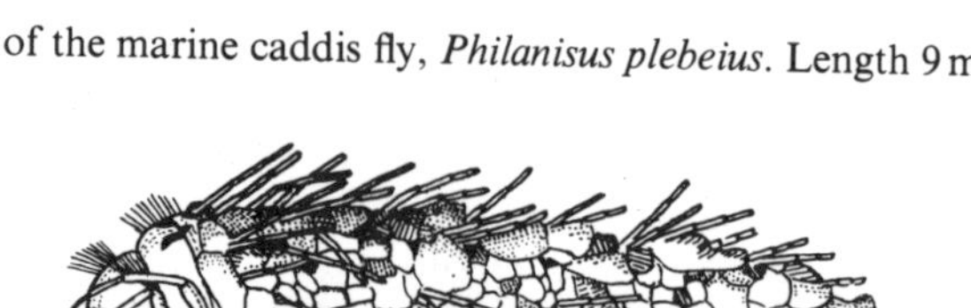

Larva of the marine caddis fly, *Philanisus plebeius*. Length 9 mm.

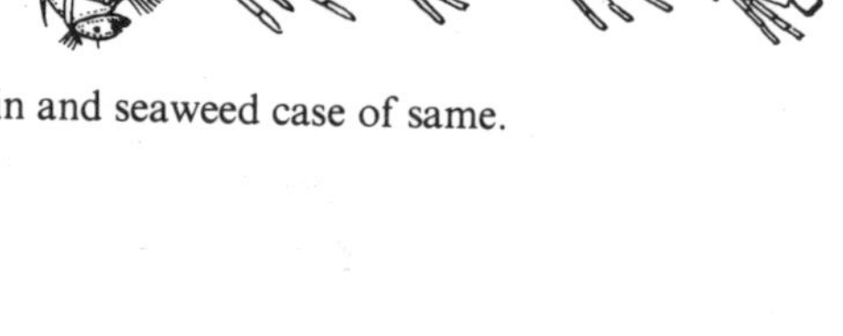

Sand grain and seaweed case of same.

caterpillars, although lacking the abdominal claspers. There are two prolegs at the back and these are hooked to anchor the larva in its case. Some larvae are predators of other aquatic insects but others live only on underwater vegetation.

Cases are made of spun silk and 'decorated' with sand grains, pebbles and plant tissues. Some species also spin a net at the entrance of their case to catch prey, while others may make their silk inside a hollow reed or other plant stem.

Larvae may sometimes be found underneath logs or stones, whereas others may be exposed on sand or gravel. They seem to prefer swift-flowing mountain streams, although some are found only in ponds. When collected, both insects and cases (or cocoons of pupae) should be retained as aids to identification.

Further Reading

Insects in General

Bastin, H. *Freaks and Marvels of Insect Life*, Hutchinson, London 1954.

CSIRO 'Scientific and Common Names of Insects and Allied Forms Occurring in Australia' (Ed. F. J. Gay) in *CSIRO Bulletin* No. 285, CSIRO, Melbourne 1966.

CSIRO *The Insects of Australia* (Ed. D. F. Waterhouse), Melbourne University Press, 1970, Supplement 1974.

Dalgleish, E. F. *Name This Insect*, The Country Book Club, London 1954.

Jeannel, R. *Introduction to Entomology* (Trans. H. Oldroyd), Hutchinson, London 1960.

McKeown, K. *Australian Insects: an Introductory Handbook*, Royal Zoological Society of N.S.W., Sydney 1942.

Norris, K. *The Collection and Preservation of Insects*, Handbook No. 1, The Australian Entomological Society, University of Queensland 1966. (This handbook is obtainable from the Department of Entomology, University of Queensland, St Lucia, Queensland 4067.

Oldroyd, H. *Collecting, Preserving and Studying Insects*, Hutchinson, London 1958.

Tillyard, R. J. *The Insects of Australia and New Zealand*, Angus and Robertson, Sydney 1926.

Williams, W. *Australian Freshwater Life: the invertebrates of Australian inland waters*, Sun Books, Melbourne 1968.

Moths and Butterflies

Burns, A. and Rotherham, E. *Australian Butterflies in Colour*, A. W. Reed, Sydney 1969.

Common, I. F. B. *Australian Moths*, Jacaranda Press, Brisbane 1963.

——— *Australian Butterflies*, Jacaranda Press, Brisbane 1964.

D'Abrera, B. *Butterflies of the Australian Region*, Lansdowne Press, Melbourne 1971.

Waterhouse, G. and Lyell, G. *The Butterflies of Australia*, Angus and Robertson, Sydney 1914.
Termites
Ratcliffe, F., Gay, F. and Greaves, T. *Australian Termites*, CSIRO, Melbourne 1952.

Index